Astronomer's Pocket Field Guide

For other titles published in this series, go to
www.springer.com/series/7814

Stefan Rumistrzewicz

A Visual Astronomer's Photographic Guide to the Deep Sky

A Pocket Field Guide

Springer

Stefan Rumistrzewicz
South Bedfordshire
UK
stefrum1974@hotmail.com

ISBN 978-1-4419-7241-5 e-ISBN 978-1-4419-7242-2
DOI 10.1007/978-1-4419-7242-2
Springer New York Dordrecht Heidelberg London

Library of Congress Control Number: 2010938908

Printed on acid-free paper

Springer is part of Springer Science+Business Media (www.springer.com)

For Theo and Charlie from your ever-loving daddy

Preface

Over the last 15 years or so there has been a huge increase in the popularity of astrophotography with the advent of digital SLR cameras and CCD imagers. These have enabled astronomers to take many images and, indeed, check images as they scan the skies. Processing techniques using computer software have also made 'developing' these images more accessible to those of us who are 'chemically challenged!' And let's face it – some of the pictures you see these days in magazines, books, and on popular web forums are, frankly, amazing!

So, why bother looking through the eyepiece you ask? Well, for one thing, setting up the equipment is quicker. You just take your 'scope out of the garage or, if you're lucky enough to own one, open the roof of your observatory, align the 'scope and off you go. If you have an equatorial mount, you'll still need to roughly polar align, but this really takes only a few moments. The 'imager' would most likely need to spend more time setting up. This would include very accurate polar alignment (for equatorial mounts), then finding a guide star using his or her finder, checking the software is functioning properly, and continuous monitoring to make sure the alignment is absolutely precise throughout the imaging run. That said, an imager with a snug 'obsy' at the end of the garden will have a quicker time setting up, but then again so will the 'visual' observer.

Another reason to look through the eyepiece is to challenge yourself. What can you *really* see? If you look long enough at an object, you'll often surprise yourself at how much visual detail you can make out. But we'll have more on observing techniques later.

Also the accessories you will need to observe and record your visual observations will cost you only around 0.5% (set of pencils, eraser, paper, and your eye!) of the cheapest imaging equipment! Money talks!

OK, but how do you record these observations? My imaging friends, you say, have a lovely portfolio/gallery of beautiful images. What will I have? Well, as mentioned above, there are ways in which you can record your observations. You can have an 'observing log' for which you will develop your own jargon (and adopt other people's) and sketches of objects of interest. Now the word 'sketching' can be interpreted in many ways. Perhaps you would never consider yourself an artist. However, after having read some books on how to sketch DSOs (see the Bibliography and Acknowledgments in this book) and having looked at some examples on line, you might well decide to give it a go. It's surprising how accurate a rendition of what you have seen at the eyepiece you can reproduce in your sketchbook, given the right tools and a little practice.

This book is not aimed at any particular level of observing 'ability', i.e., beginner, intermediate, or expert. We will assume that you will be able, perhaps with a little ingenuity, to locate the DSOs mentioned in this book using a GO-TO system or an accurate star chart. There are many of the latter available (look in the appendices of this book for help there). However, this book will help to ensure that you are looking at the correct object, and yes, it really is that faint! This book should be helpful to those of us that are new to DSO astronomy and are expecting to see the fantastic bright and colorful images one sees in magazines but miss the faint fuzzy that is in the middle of the field of view!

The images you'll find in this book are taken using a Watec 120n video camera. This camera is able to give 'live' views of DSOs when placed in an eyepiece and seemed the most suitable camera to use in order to give an accurate representation of what the suburban observer may actually see. All the images are, therefore, single frames and completely unprocessed. The only exception is where an odd label has been added here and there to distinguish between several objects in the same field of view. Also included are some sketches, which have been scanned onto computer and the colors merely inverted to give a white-on-black view - i.e., no touching up! Where possible, views of the object are shown through two different-sized telescopes - either through a 72mm aperture refractor or an 11 inch SCT.

As you can appreciate from the title, this book is about challenging yourself to 'see more.' It's not about how to find a DSO, although it can be used to confirm that what you are actually observing is the correct target! You will not find any colorful Hubble Space Telescope images within these pages. This book is all about what you can actually see, what a DSO really looks like under suburban (read 'orange') skies. Often a casual look will leave you disappointed, especially when you've seen the beautiful pictures of the same object in a magazine or online. But you'll be surprised what you can see in your own backyard if you take your time to study that faint smudge.

As you will see, when you reach the relevant section, the DSOs in each constellation are listed in terms of object type - i.e., 'Planetary Nebula' - and in the order of catalog, that is to say: Caldwell, IC, Messier, and NGC. The list in each constellation is by no means exhaustive, as this book is only really intended to get you started (or, perhaps, re-started) on the road to visual astronomy and sketching. The idea is to sketch your way through this book and/or challenge yourself to see the extra detail mentioned with some of the objects. So good luck, clear skies, and enjoy!

South Bedfordshire, UK Stefan Rumistrzewicz

Acknowledgments

Firstly, I would very much like to thank Bern from *Modern Astronomy* for his excellent advice in the purchase of an appropriate camera with which to show eyepiece views. I would also like to thank all those that have given me help and advice on the excellent astronomy web forum, Stargazers Lounge (SGL). Also, related to SGL, I would like to thank the forum's sponsor, First Light Optics, for their friendly advice and support when requested. My thanks must also go to Simon from the Widescreen Centre in London for his sound advice in my first telescope purchase since returning to the UK.

I would also like to acknowledge my brother-in-law, Julian, without whose help the trolley and ramp would not have been built, and there would certainly not be as many photographs in this book. Many thanks, Jules!

Last, but by no means least, I must thank my ever-understanding and supportive wife, who never ceases to amaze me in her indulgence of my hobby.

Contents

About the Author

My love affair with the cosmos began, probably, when I stared, open-jawed, for two hours at the film *Star Wars*. Although I was only three at the time, I can remember subsequently drawing pictures of stars and planets and 'X-wings.' I quickly became a science fiction addict, watching re-runs of *Star Trek* and various other series. And so, I desperately wanted a way of seeing some of these objects for myself. My parents bought me a cheap (supermarket) 'scope when I was 11, which I then proceeded to set up in my sister's room (the only one with a southerly view) and stargaze through her window. Neither my father nor I could ever get it to focus properly. Of course, at the time, I had no knowledge or understanding of air currents, seeing, or indeed how to use a telescope or star chart. Nevertheless, even with the horrendous optics of this small telescope I could make out some craters on the Moon – I was hooked!

I decided to 'save up' for something better. After several birthdays I finally bought myself a reasonable 'scope – a 9.25 inch Dobsonian, and spent many a clear night sitting on the garage roof (my sister wanted her room back!) gazing at the wonders of the universe.

Soon after, university beckoned. I studied Chemistry, with a subsidiary in Astronomy. Although I very much enjoyed the latter, I was unable to continue as the university (at the time) did not offer Astronomy as a major degree. And so, my hobby took an enforced break, which was then extended by teacher training, getting married, and emigrating to east Africa. It was there, under the clear equatorial skies of Kenya, that my love of the cosmos was rekindled and my wife bought me a 5″ Newtonian telescope for my birthday. I was astonished by the things I was able to see with such a small aperture. It just goes to show what difference a truly dark sky makes.

Since then, I have been interested in visual astronomy and what deep sky objects really look like. Having subsequently returned to the UK and settled in South Bedfordshire, I developed a keen interest in sketching at the eyepiece and have spent many a nocturnal hour gazing through my 'scope, having told my wife I wouldn't be outside for 'too long.'

Eventually, I began to wonder why there wasn't a book that showed photographs of the 'visual' image (i.e., what can be seen at the eyepiece). It seemed that descriptions of DSOs varied slightly from writer to writer and Hubble

Space Telescope images told me very little about what I could actually see in my backyard. That was how this book was conceived. And so now, here it is… I trust that it will give you as much pleasure to read as it was for me to write.

Enjoy and clear skies!

CHAPTER 1

OBSERVING PLANS AND TECHNIQUES

Before You Start

Having a plan or an idea as to what you wish to observe can save you a lot of time fiddling around in the dark. By looking at a star chart or computer software one can devise one's own observing program for an evening's viewing. Alternatively, there are a number of books available that contain suggested sky tours and interesting observing programs. Having such a plan can save you a lot of time on the night. It can be very frustrating, having set up for a night's viewing, to have spent half an hour deciding what to look at, only to look up and see the clouds roll in!

As an astronomer you are probably very meteorologically aware. After looking at the weather forecast on the television or the Internet, check any satellite footage. This is invaluable for us astronomers and is an area where the Internet can be extremely useful. There are several sites that give regularly updated satellite images, examples of which include http://www.sat24.com for Europe and Africa or http://www.intellicast.com/National/Satellite for US images. Try out the 'Pan and Zoom' feature! Particularly useful are the infrared images, as they can be seen at night.

If your telescope is stored in a garage, for example (see below), then setting it up before you go inside after arriving home from work is a great way of maximizing observation time. Your telescope is already cooling down to ambient temperatures while you are greeting the family or loved ones after a hard day's work. This is a particularly useful timesaver for cooling the optics of a large SCT, which can take up to two hours.

What about storing the telescope outside? By this we don't for a minute mean that you should leave your pride and joy nicely set up in the middle of the garden in all kinds of weather and, indeed where someone might be tempted to relieve you of it in your absence. No. By storing outside we only mean to store it outside the house or your apartment, preferably in a

S. Rumistrzewicz, *A Visual Astronomer's Photographic Guide to the Deep Sky*, Astronomer's Pocket Field Guide, DOI 10.1007/978-1-4419-7242-2_1,

secure shed or garage. As these are not, usually, heated, storing your telescope thus will dramatically reduce the amount of time you will spend waiting for the optics to cool down to ambient temperatures outdoors.

An observatory is the ultimate in time saving. That small dome or, indeed, roll-off roof shed at the bottom of the garden is a fantastic place to retreat to after a hard day's work. Open the roof, flick the switches, and off you go! With this setup, you have very little waiting time for your telescope's optics to cool. If the clouds roll in and rain threatens, you just close the roof. You can even treat yourself to a 'quick astro-fix' if there is a brief gap in the clouds after dinner. Of course, building and maintaining the 'obsy,' as it is

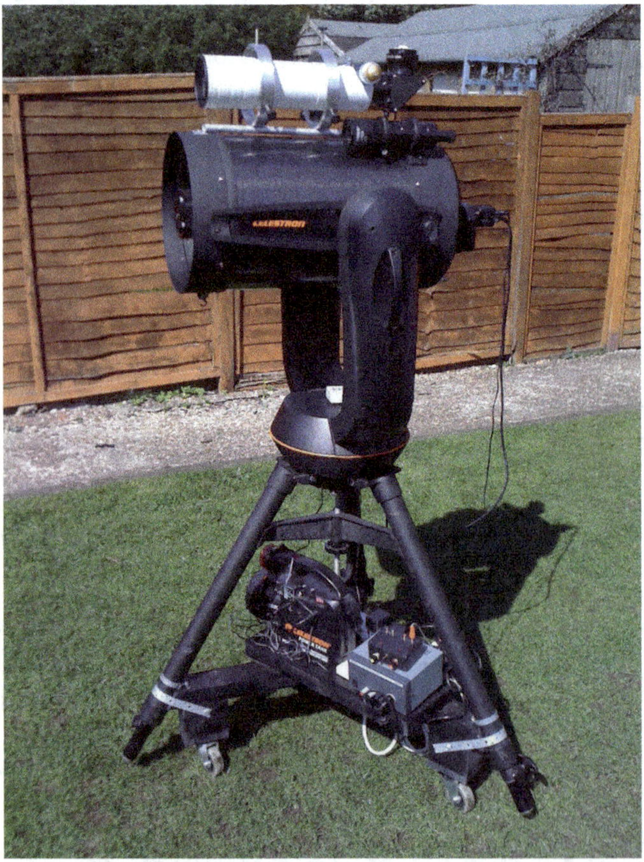

affectionately known, is another matter and will require a little DIY skill, unless you have a large cash reserve. However, there are less expensive alternatives, especially for those of us whose gardens are little bigger than a postage stamp!

The building of an observatory will not be discussed here, but there are several books available on the subject, two from this publisher: Arditti's *Setting up a Small Observatory* and Hick's *Building a Roll-Off Roof Observatory.*

Trolleys can be very useful in speeding up set up time for those of us without an obsy. The telescope is set up on the trolley. This is then just 'wheeled out' from the garage or shed when required. The author recently (with help) designed and built a trolley for an 11″ SCT. This included a rather handy tray upon which are all the necessary power packs, dew heater controllers etc. As the garden is too small to hold an observatory, this simple device saves a considerable amount of set up time. A picture of the setup is above.

For those of us who are DIY-challenged, one can purchase trolleys suitable for one's telescope. Rather bizarrely, at the time of writing, there were very few companies that had spotted this gap in the market. However, one such company is, http://www.scopebuggy.com.

Observing Techniques and Tips

Averted Vision

This often talked about technique is perhaps the most powerful tool in the 'Mark One' observer's arsenal. It really does work! The scientific principle behind it is very straightforward. The most optically sensitive part of the retina is actually off centered. The retina is covered in two types of light-detecting cells called rods and cones (due to their shapes). The cones are for color vision, whereas the rods are for low light vision. Your retina has about 20 times more rod cells than cones; in fact it has about 100 million of these cells! The cones are located fairly centrally, but the rods are spread over a larger area. Therefore, by not looking directly at the object, but just to the side of it, a wealth of extra detail often comes into view as the faint light from the object impinges on the most sensitive part of the retina, i.e., where the rods are. This technique is particularly useful when observing nebulae and galaxies. It is surprising how much extra detail can be gleaned from the faintest of galaxies (M33 for example) by just using averted vision and studying the region for several minutes.

Movement

Your eyes are very sensitive to movement. Sometimes by simply tapping the OTA, certain objects become fleetingly visible. Again this is very useful for dim DSOs. Combining this with the above technique can bring you ever closer to your telescope's theoretical magnitude limit.

Time

This may seem a strange 'technique' to include here. However, taking your time to really look at an object, brings out a remarkable amount of detail. As mentioned earlier, a casual glance at a faint fuzzy can be very disappointing. But spending more time, using the techniques described above, will often bear fruit – possibly bringing that elusive halo, or those faint spiral arms, into view.

Also time, in terms of experience is useful. Coming back to an object you first observed some time ago, to see if you can glean more detail now that your eyes know what to look for can yield surprising results. This is where accurate records of your observations and/or sketches are a great record one can use for comparison.

CHAPTER 2

ACCESSORIES AND 'PIMPING' YOUR 'SCOPE

Dew Shields

SCTs, Maksutov-Newtonians and refractors are dew magnets. The dew shield is an absolute must to keep this enemy at bay. Dew heaters can also be very useful for this purpose. However, the dew shield can also serve another purpose and not just for these telescope types but, in fact, for all telescope types. Because the shield extends beyond the end of the OTA, it is able to prevent stray light, say from a street light, from entering directly into the telescope. That is to say, the shield casts a shadow over the end of the OTA. This means that the light from the DSO is not interfered with by any other light source.

Head Shroud

Placing a hood or shroud over your head as you observe through the eyepiece is a great way of preventing stray light from getting between you and the view! Yes, you do look a little like someone taking a photo in the early part of the last century, but in your backyard no one will see you, so give it a try. Again, this greatly helps you to get more contrast.

Flocking

All surfaces reflect light to a certain degree. The smoother the surface, the better it is at reflecting light. This is what you want for your primary and secondary mirrors in a reflecting telescope. However, if you don't want stray light from your neighbor's security system bouncing around inside your OTA and thence onto your mirrors, you need to make the inside of your tube as rough as possible.

S. Rumistrzewicz, *A Visual Astronomer's Photographic Guide to the Deep Sky*, Astronomer's Pocket Field Guide, DOI 10.1007/978-1-4419-7242-2_2,

There are several ways of doing this. Some (usually more expensive) tubes are 'baffled.' This is a system of grooves or rivets that run perpendicular to the inner surface of the OTA and along its circumference. This makes it very difficult for stray light, entering at angles from the next garden, say, to reflect itself along the length of the tube and into the eyepiece.

Another way is to use a special non-reflective paint, again reducing the reflectivity of the inner surfaces.

Flocking works similarly to the latter method. It requires either special flocking material or, more simply, some felt and sticky tape. Simply stick the material to the inside of the OTA, particularly behind the secondary mirror. Be careful not to drop any material onto your mirrors!

A combination of this and a dew shield will greatly increase contrast and improve your viewing experience. It is not uncommon that a combination of flocking, dew shield, and head shroud can lower the magnitude of what you can see by up to half a magnitude. That is to say, if you could only previously see down to 11mag in your telescope, you may now be able to see down to 11.5mag – bringing many more objects into view. However, it should be noted that none of these methods will enable you to see beyond the theoretical limits of your telescope. These limits are determined by the aperture and the type of 'scope you have, i.e., refractor or reflector. Reflectors have a secondary mirror that obviously cuts out some of the incoming light, and so the effective aperture is slightly decreased.

Filters

Astronomical filters work very simply by blocking certain frequencies of light, while allowing others to pass through undeterred. There are a huge number of varieties of filter available for both visual and photographic astronomers. For the purposes of this book, we will only discuss visual filters.

For astronomers in the suburbs, as most of us are, the scourge of light pollution is ever present. This, as you are no doubt aware, makes our night skies orange rather than the beautiful blue/black they should be. This ambient light effectively overpowers the faint light coming from DSOs, making many practically invisible, even in the largest telescopes. However, all is not lost.

By isolating, i.e., cutting out, the frequencies of light generated by street lights (the most common source of ambient light), we can allow other frequencies, i.e., those from DSOs, to come through. This is where filters come into play. Nevertheless, some of the frequencies of light generated by DSOs (depending on which sort you look at) are the same as those of ambient light, so although the filter will improve the contrast between the object and the background sky, it will also make the object slightly dimmer.

First up is the Neodymium filter. This filter is probably the cheapest type of contrast filter available. It reduces some of the background sky glow, making the sky appear slightly darker, without reducing the light from the object being observed. It can be classified as a multi-purpose filter in the sense that it can be used for all types of DSOs. The effects are subtle but nevertheless distinct, enabling you to detect, for example, faint wisps of nebulosity in contrast with the background sky. Astronomers have also reported that this type of filter also helps to add 'crispness' to planetary and lunar views.

Next are the broadband filters, the UHC (ultra-high contrast) filters, sometimes referred to as anti-light pollution (ALP or LPR) filters. These types of filters, generally speaking, cut out the frequencies generated by mercury and sodium streetlights. In suburban and urban areas, this type of filter is extremely useful, as it really does darken the background skies quite significantly. However, there is a tradeoff with this seemingly miraculous solution to our suburban woes. As mentioned earlier in this section, certain DSOs emit light over a large range of frequencies, including those of sodium and mercury. These DSOs are, of course, those that involve starlight, i.e., galaxies and clusters. Therefore, observing a galaxy using an UHC filter may help to distinguish it from the background sky, but paradoxically it will make it appear much dimmer and possibly invisible!

Finally for the visual astronomer we have the narrowband filters, such as the OIII (doubly ionized oxygen) and H-Beta (doubly ionized hydrogen). As you can imagine from their name, these filters only allow very specific frequencies of light through. In the case of OIII, usually only the frequencies associated with this state of oxygen (doubly ionized) will pass through to your eye. This filter is extremely useful for emission nebulae and planetary nebulae, as the contrast with the background sky is quite dramatic. Try it for yourself on the Veil Nebula in Cygnus, for example. It will also have the effect of blocking a lot of starlight, so stars in the field of view of your desired object may disappear from view altogether! The H-Beta filter has a similar effect in terms of contrast, but far fewer DSOs emit at this frequency in significant amounts. It is, however, extremely useful for spotting the Horsehead Nebula in Orion. For this reason, H-Beta filters are often referred to as Horsehead filters.

The following images show the effects of these filters on an emission nebula, NGC 2359, also known as Thor's Helmet.

Note that in the first image shown, the nebulosity is barely visible and only with averted vision in a few places around the small grouping of stars.

The effect of the LPR filter on the visibility of the nebulosity is more apparent in the next image. More nebulosity has now come into view, as the OIII emissions are slightly enhanced and mercury emissions (from street lights) are suppressed.

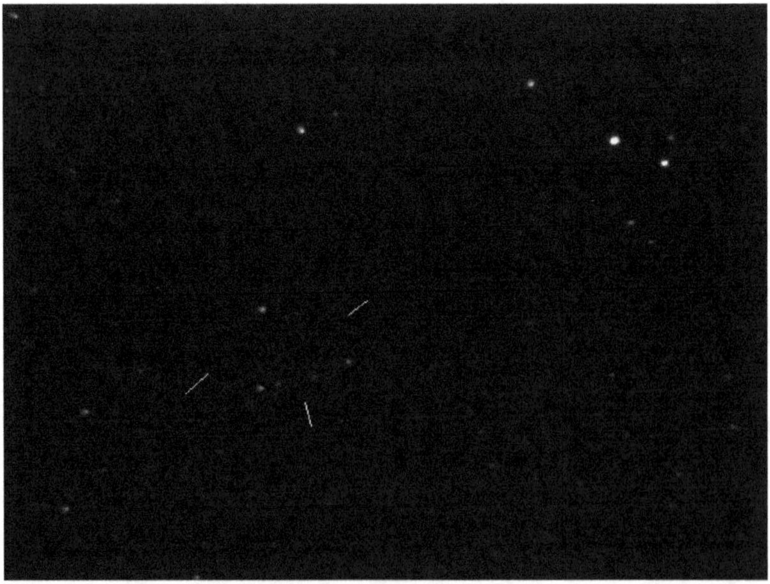

NGC 2359 11" SCT using a Neodymium filter

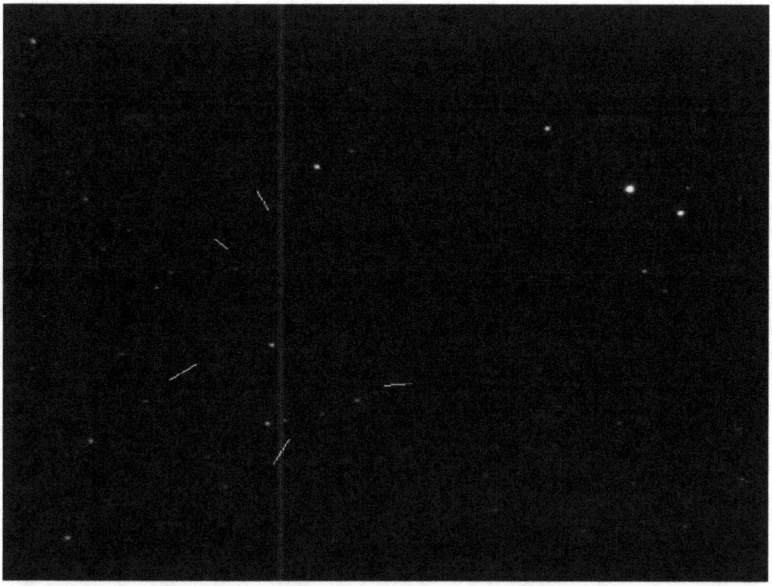

NGC 2359 11" SCT using a LPR filter

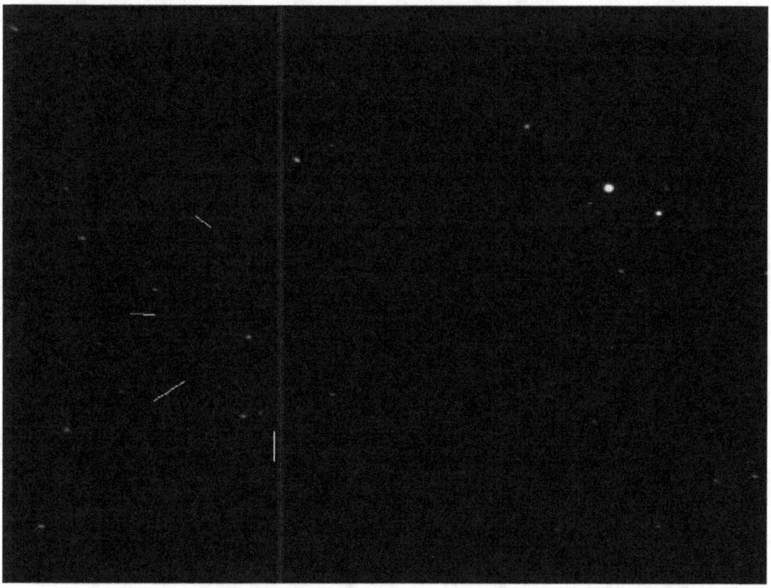

NGC 2359 11" SCT using an OIII filter

The OIII filter has further enhanced the nebulosity by blocking out other frequencies of light. Note that now fewer stars are visible in the field. More faint nebulosity has come into view in the top left-hand corner of the image and some around the bright star in the top right-hand corner.

CHAPTER 3

SKETCHING

This aspect of the hobby has undergone a resurgence in recent years. Although there is, of course, a huge proliferation of relatively cheap CCD cameras for the budding astro-imager, astronomers are coming back to using the ol' mark one and a pencil. Perhaps because it is cheaper and quicker to set up. But also, it requires you to really study the object if you're going to try and sketch it.

Perhaps more remarkably, the artistic skills required to make a reasonable impression of what you have seen are very small and need only a little practice. Of course, as with anything, practice really does make perfect, and the more sketches you do, the better you will become! To get an idea of what you might achieve, there are several very talented astro-sketchers around, whose handiwork can be seen on the Internet, on forums such as Stargazerslounge or the Astronomy Sketching Forum (on astronomyforum.net), or in magazines such as *Sky & Telescope, Sky at Night, Astronomy,* and *Astronomy Now*. There are also some books available on this subject. A recent one from this publisher is *Astronomical Sketching: A Step by Step Introduction* by Handy, Moody, Perez, Rix, and Robbins.

For your interest we have included a couple of this author's efforts below. These were done using a pencil and sketching paper. The drawing was scanned (so that you can see it here!) and then the colors inverted.

S. Rumistrzewicz, *A Visual Astronomer's Photographic Guide to the Deep Sky*, Astronomer's Pocket Field Guide, DOI 10.1007/978-1-4419-7242-2_3,

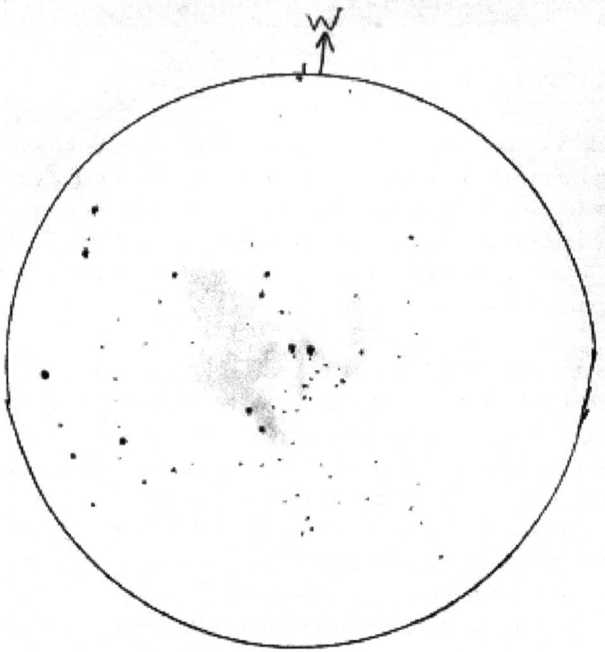

M16 & IC 4703 – The Eagle Nebula (using an 11″ SCT and UHC filter)

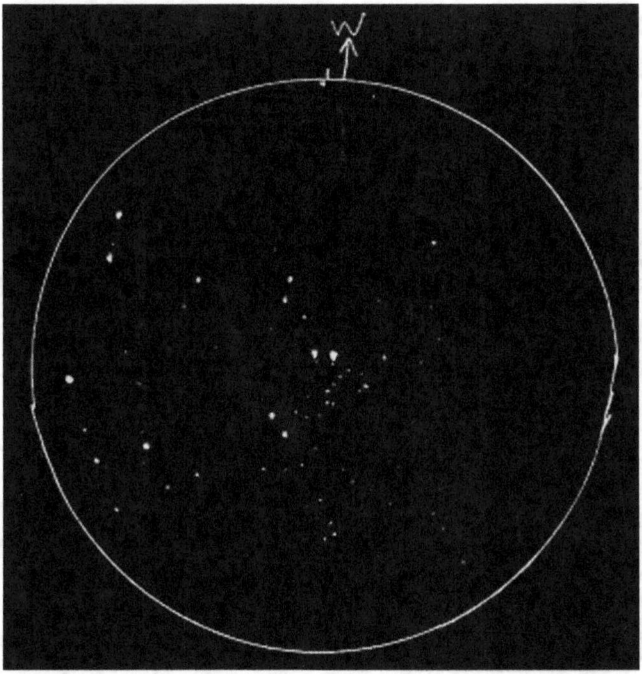

M16 and IC 4703, colors inverted

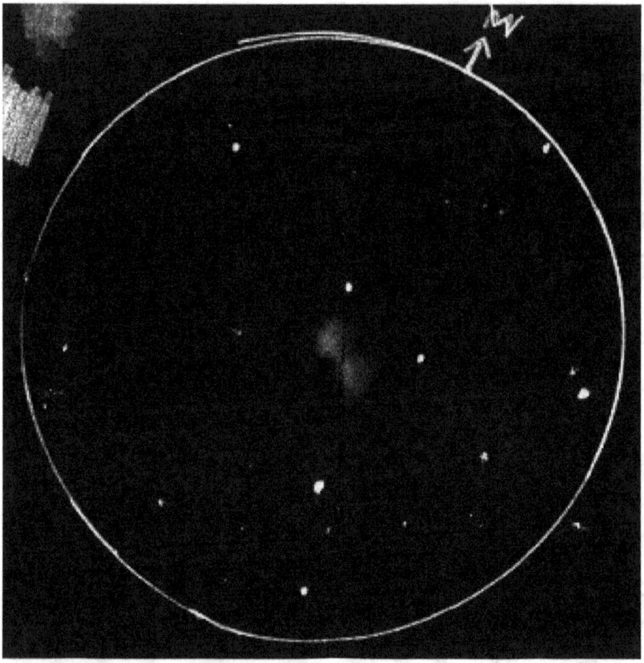

M76 – The Little Dumbbell (using an 11" SCT)

CHAPTER 4

CONSTELLATION OBSERVING LISTS AND PHOTOS

The rest of this book contains a series of sections listed in constellation alphabetical order, i.e., Andromeda first, Vulpecula last. Each constellation begins with a list of deep sky objects, the object type (i.e., galaxy, cluster, etc.), their positions on the celestial sphere, their approximate size, and, where appropriate, their surface brightness.

For each object we have indicated what type of object it is, using a system of abbreviations (except for galaxies just listed as 'Galaxy'), as follows:

EN Emission nebula

RN Reflection nebula

SR Supernova remnant

OC Open cluster

GC Globular cluster

SC Star cloud

CN Cluster with nebulosity

DS Double star

The positions for each object are given using the International Celestial Reference System in J2000.0 coordinates. Although many of you may have a GO-TO system that will include the vast majority of the objects in this book, it was felt that positional information would be useful for those of you who use digital setting circles.

S. Rumistrzewicz, *A Visual Astronomer's Photographic Guide to the Deep Sky*,
Astronomer's Pocket Field Guide, DOI 10.1007/978-1-4419-7242-2_4,

Also included is the approximate size of each object, so as to give you, the viewer, some idea of how big the object should appear in your eyepiece's field of view. If the object appears much smaller – and galaxies often do – then you can challenge yourself to see if you can pick out that extra, very dim, piece of detail!

As mentioned previously, wherever possible, we have indicated the approximate surface brightness of the object. This is to give you an idea as to whether the object will be at all visible in your setup, but also to enable you to give yourself 'a pat on the back' if you manage to see an object at the very limit of your telescope's light grasp.

Please note that the object list for each constellation is by no means exhaustive and, in some constellations, lacking in certain bright objects due to their southerly declination. As this book is aimed at observers in the northern hemisphere, most of the southern constellations are not included. No objects below a declination of −21° are included due to the difficulty in obtaining reasonable pictures from the author's backyard location (as looking south meant looking through light polluted murk and over the author's house!).

As was mentioned before, each object has a very brief description and accompanying photo showing what you will see at the eyepiece using that sized 'scope from a suburban location. Where practicable, there are comparative photos taken using either an 11″ SCT or a 72mm refractor. All the photos show the unfiltered view. The field of view with the 11″ SCT is approximately 23′×17′, whereas the 72mm refractor is approximately 45′×34′. Some of the objects listed for a constellation will include challenges – i.e., can you spot the extra detail? Some descriptions also include tips, for example: which is the most appropriate filter to use. Where objects are particularly faint, or difficult to distinguish from the field, white marker lines have been included. For some of these objects the reader may need to use averted vision to see them on the page! Please note that in the 11″ SCT photos, the orientation is north down. The orientation of the refractor photos is north up, and east and west are reversed.

There are visual descriptions for a handful of objects for which the author could not obtain adequate photos. In these cases, it is either because the object's angular size was so large that it could not be shown adequately using the equipment or the object is plainly visible using filtration (i.e., NGC 7293, the Helix Nebul*a*) but not unfiltered. All photos in this book are supposed to represent the unfiltered view!

You are also invited to try and use the Trumpler classification system to classify open clusters for yourself (see below). The accepted (modern) answers are given at the bottom of their respective page – don't cheat! However, there can be some discrepancy between different sources. In these cases both are listed, i.e., NGC 2281 is I 3 r/m.

Trumpler Classification System for Open Clusters

This is a relatively simple classification system, based on how detached and concentrated the open cluster is, the range of brightness of the constituent stars, and the approximate number of stars (richness). The system is given as follows:

Concentration

I. Detached from the field with a strong concentration towards the center

II. Detached from the field with a weak concentration towards the center

III. Detached but no concentration towards the center

IV. Difficult to detach from the surrounding star field

Range of Brightness

1. Small range of brightness

2. Medium range of brightness

3. Large range of brightness

Richness

p. Poor (less than 50 stars)

m. Moderately rich (between 50 and 100 stars)

r. Rich (more than 100 stars)

So, when classifying an open cluster, one lists these three values. For example:

M 37 (in Auriga) is classified as I 2 r. This means it is well detached from the star field and clearly distinguishable. It is very concentrated towards the center. There is a definite range of brightness among the stars within the cluster, and there are over 100 stars in it!

NGC 6709 (in Aquila) is classified as IV 2 m. This means that it is very difficult to distinguish from the surrounding star field and shows no concentration towards the center. The stars show a range of brightness, and there are between 50 and 100 cluster members.

Shapley–Sawyer Concentration Class for Globular Clusters

Although not given in the following pages, this classification system is used for globular clusters (GC). Roman numerals are used, and the GCs are classified according to concentration along a 12-point scale (using Roman numerals) counting downwards. That is to say, XII is the least concentrated and I is the most. Examples would be: M 12 (in Ophiuchus) is IX; M 3 (in Canes Venatici) is VI and M 5 (in Serpens) is V.

Hubble's Galaxy Classification System

This system, affectionately referred to as 'The Tuning Fork,' was first conceived by Edwin Hubble in 1926 and has undergone several revisions since (most notably by de Vaucouleurs). Nevertheless, it still forms a basis for classification. Here is a diagram showing the famous 'tuning fork.'

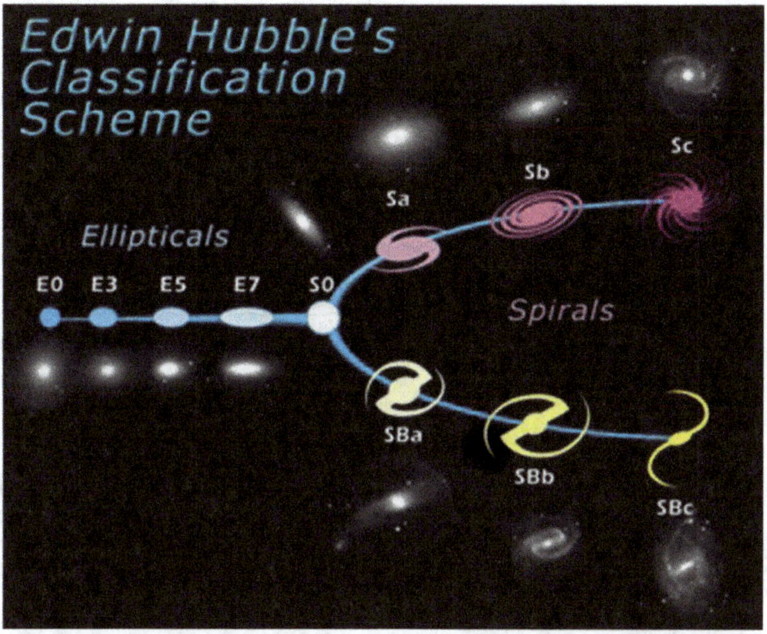

Image courtesy of NASA and ESA

The diagram above shows how galaxies were divided by Hubble into various types according to their shape. These are: elliptical (E), lenticular (S0), spiral (S), and barred spiral (SB). There is also another class, not shown on this diagram, known as 'irregular' (I).

When he devised this system, Hubble called the elliptical galaxies 'early.' Since then we have learned that the evolutionary sequence for galaxies does not follow this path. Nevertheless, it is still a useful system of classification and has formed the foundations of the systems used by astronomers today. Note that the classification system used in this book is the de Vaucouleurs classification.

For further reading and a more in depth explanation of galaxy classification systems refer to Steinicke and Jakiel's excellent book *Galaxies and How to Observe Them* (see Recommended Texts), available from this publisher.

Have fun and see what you can see!

Andromeda

Objects	Type	Position	Size	SB*
M 31	Galaxy	0h42′33.3″ +41 15 20	190.5×61.7	13.5
M 32	Galaxy	0h42′41.7″ +40 51 59	8.7×6.5	12.5
M 110	Galaxy	0h40′19.5″ +41 41 25	21.9×11	14
NGC 404	Galaxy	0h09′27.0″ +35 43 02	3.5×3.5	12.9
NGC 752	OC	1h57′35.0″ +37 50 00	75	
NGC 891	Galaxy	2h22′33.8″ +42 21 01	11.7×1.6	13
NGC 7662	PN	23h25′53.9″ +42 32 06	37″	

M 31

This is our famous galactic neighbor, the Andromeda Galaxy. Classified as an SAb galaxy, it is visible to the naked eye under dark skies. It requires large fields of view due to its angular size. Can you see NGC 206 (about 1° south west of the core), a cluster in the galaxy itself?

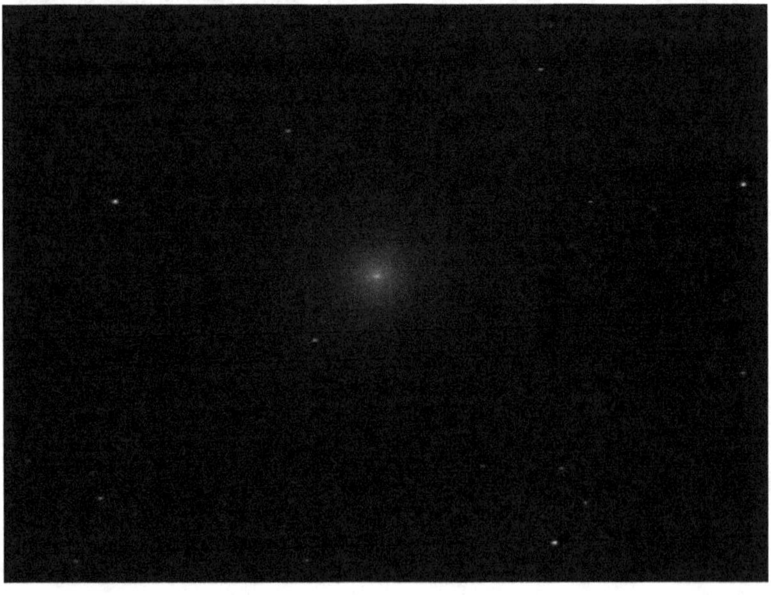

M 31 11″ SCT

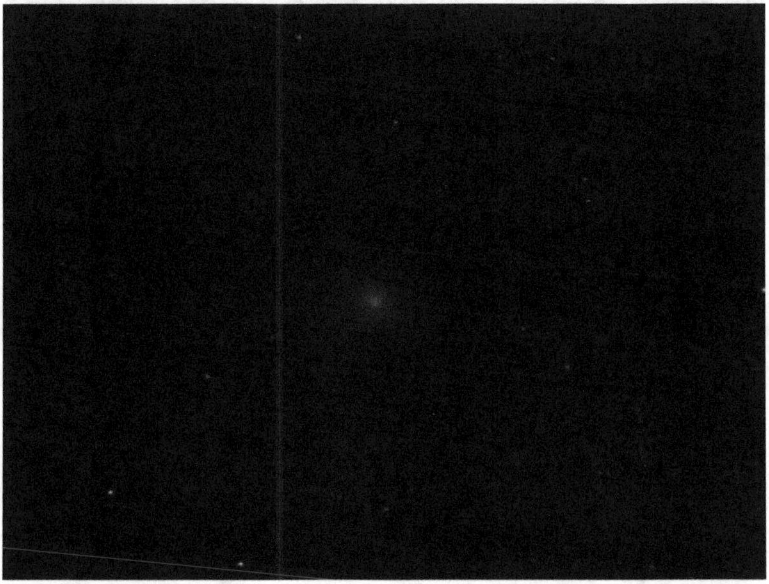

M 31 72mm Refractor

M 32

The galaxy is classed as cE2 and is one of the satellites of M 31. Note the almost stellar core.

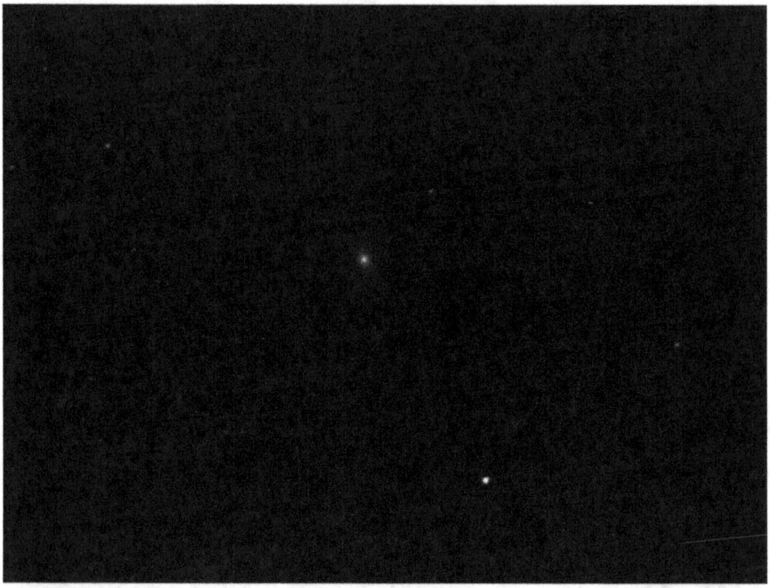

M 32 11" SCT

M 32 72mm Refractor

M 110

Faintest of the M 31 satellites, this is an E5pec galaxy with a low surface brightness and faint stellar core.

M 110 11″ SCT

M 110 72mm Refractor

NGC 404

This is an SA0 galaxy whose close proximity to the star Mirach (b Andromedae) makes it difficult to see because of the latter's glare. Try masking Mirach by taking it out of the field of view.

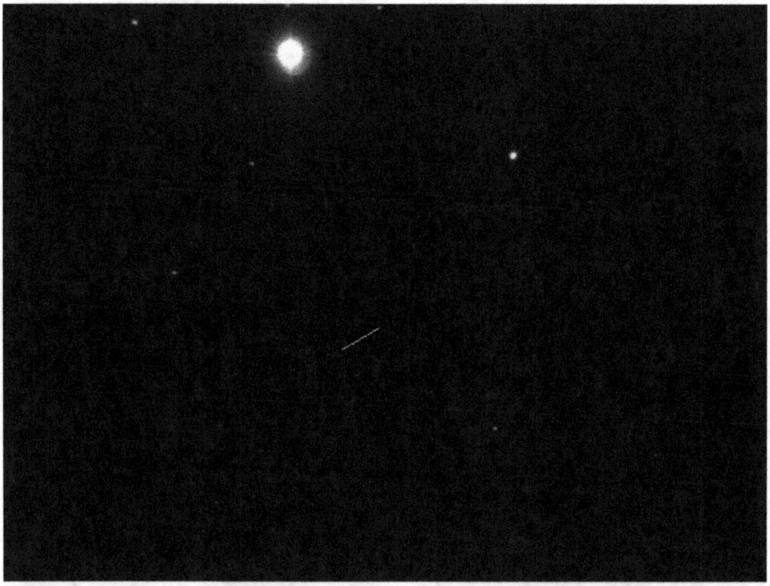

M 32 11" SCT

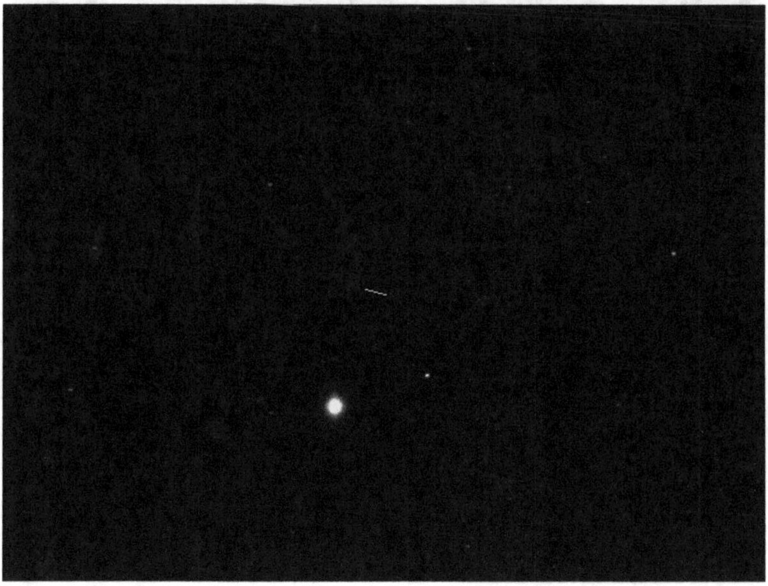

M 32 72mm Refractor

NGC 752

This open cluster contains over 80 stars and is best observed at low powers, as it requires a wide field of view.

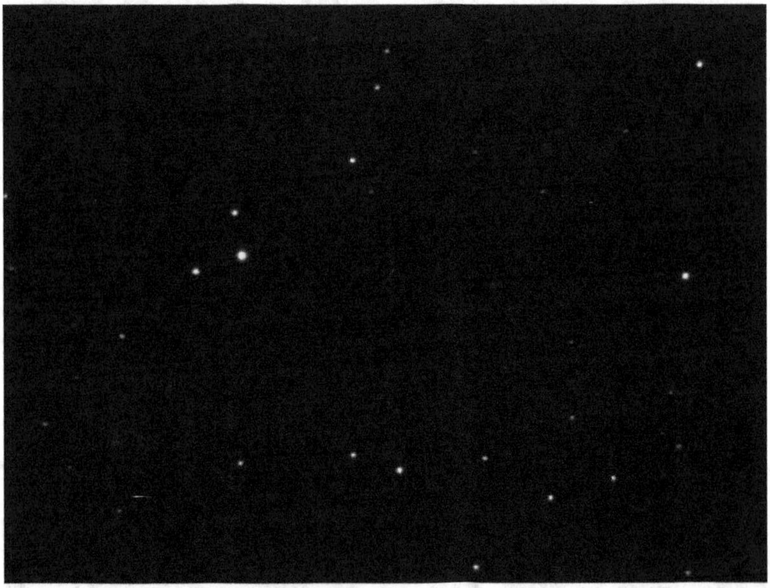

NGC 752 11″ SCT

NGC 752 72mm Refractor

Trumpler Classification: II 2 r

NGC 891

This is an SAb galaxy presented edge-on. Can you see the dark lane running along full length? If so, can you see the finger-like protrusions from it (use high magnification)?

NGC 891 11″ SCT

NGC 891 72mm Refractor

NGC 7662

This planetary nebula is better known as the Blue Snowball, due to its bluish color. Increasing the magnification to high power should reveal a slight elongation N–S and an almost rectangular bright region within the nebula. Can you see these details? Can you see the central star at mag13.3?

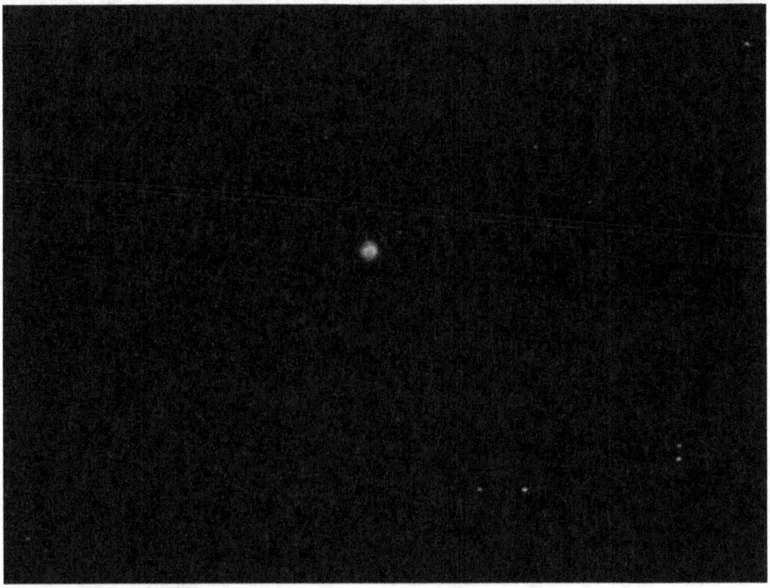

NGC 7662 11″ SCT

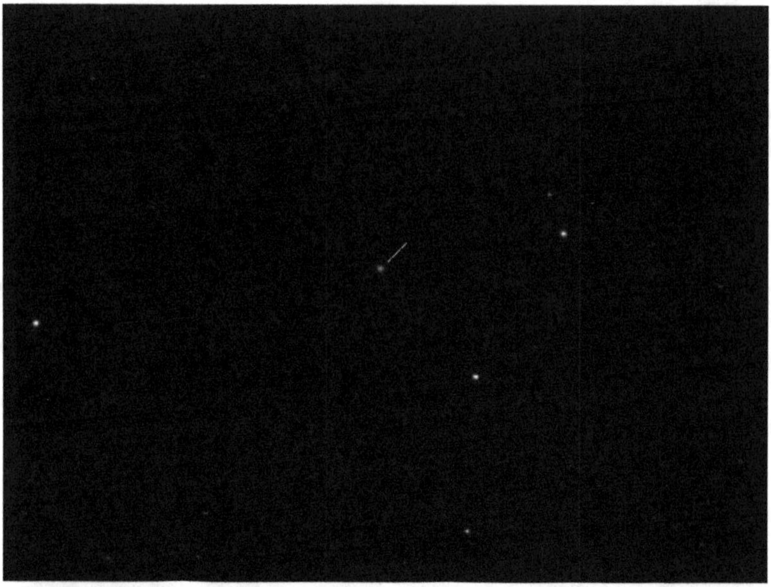

NGC 7662 72mm Refractor

Aquarius

Objects	Type	Position	Size	SB*
M 2	GC	21h33′27.0″ −00 49 12	16.0′	
M 72	GC	20h53′27.9″ −12 32 13	6.6′	
M 73	OC(?)	20h58′56.0″ −12 38 00	1.4′	
NGC 7009	PN	21h04′10.9″ −11 21 49	35″	
NGC 7293	PN	22h29′38.4″ −20 50 12	16′	

M 2

This is quite a large globular cluster and is very concentrated towards the center. Can you detect the curving dark lane in the northeast part?

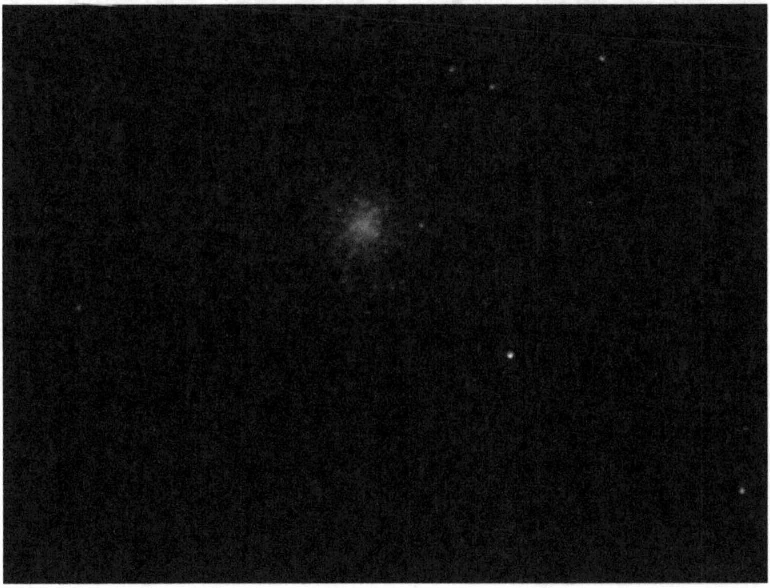

M 2 11″ SCT

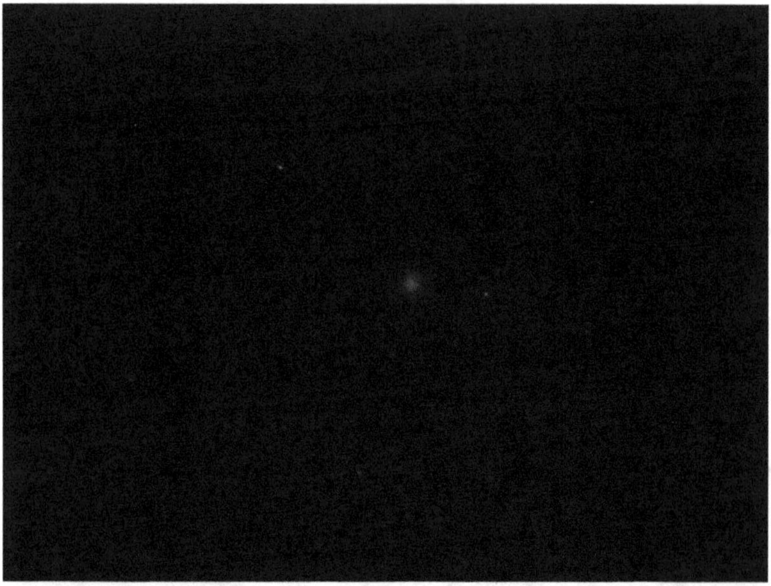

M 2 72mm Refractor

M 72

This is a fairly loose and faint globular cluster with a good deal of structure visible in the core. Use a high power and a large aperture for this one!

M 72 11" SCT

M 72 72mm Refractor

M 73

Although listed as an open cluster, this small grouping of stars may not be a true cluster, as the four main stars apparently have different proper motions. The jury is still out….

Trumpler Classification: IV 1 p (?)

M 73 11″ SCT

M 73 72mm Refractor

NGC 7009

This bright aqua-colored planetary nebula is named the Saturn Nebula, due to its oval shape and faint extensions (ansae). Can you see the faint outer nebulosity surrounding the bright oval shape (use high power)? Can you see the ansae? An UHC or an OIII filter will help.

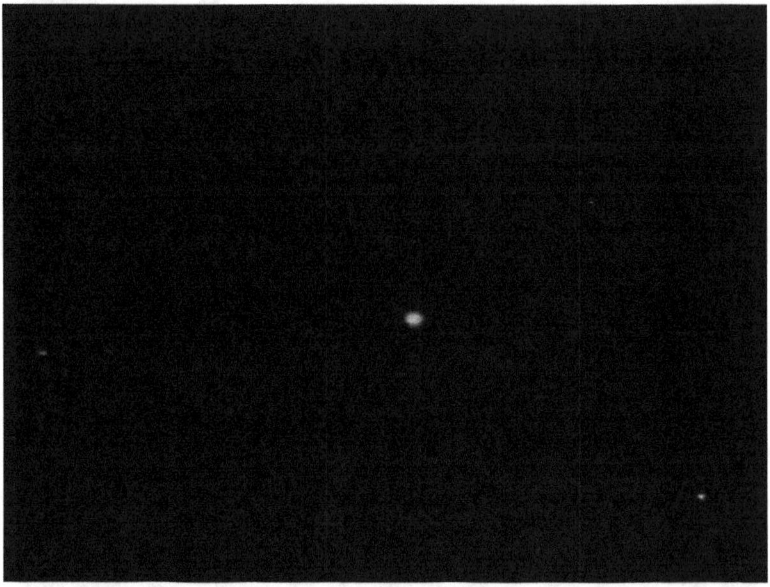

NGC 7009 11″ SCT

NGC 7009 72mm Refractor

NGC 7293

This apparently large planetary, known as the Helix Nebula, has a low surface brightness due to its size. The central ring, about 12′, should be visible using a UHC or an OIII filter to boost the contrast with the surroundings. Unfortunately this object was invisible from the author's location without filtration, as it was lost in the 'murk,' but it is included here, as you should be able to locate it even with a small telescope. With an UHC or an OIII filter, it should resemble a much larger version of the Ring Nebula, M 57, in Lyra.

Aquila

Objects	Type	Position	Size	SB*
NGC 6709	OC	18h51′30.0″ +10 20 00	15′	
NGC 6738	OC(?)	19h01′24.0″ +11 36 00	15′	
NGC 6751	PN	19h05′55.5″ −05 59 33	26″	
NGC 6755	OC	19h07′49.0″ +04 16 00	15′	
NGC 6760	GC	19h11′12.1″ +01 01 50	9.6′	
NGC 6781	PN	19h18′28.2″ +06 32 15	114″	
NGC 6804	PN	19h31′35.2″ +09 13 31	66″	

NGC 6709

This is quite irregular as OCs go, with no obvious center, but it contains around 40 stars with a moderate range of brightness.

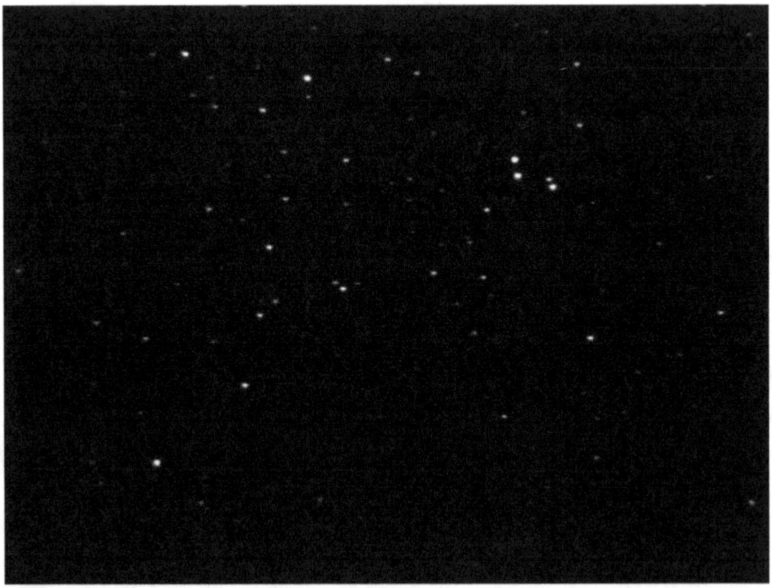

NGC 6709 11″ SCT

NGC 6709 72mm Refractor

Trumpler Classification: IV 2 m

NGC 6738

This 'open cluster' has few stars and is not well detached from the field. Recent studies have shown that this is not, in fact, a true cluster but a random grouping of bright stars!

NGC 6738 11″ SCT

NGC 6738 72mm Refractor

NGC 6751

This apparently small planetary nebula requires quite high powers and appears slightly green in color. Can you see the central star? An OIII filter helps to show some of the 'outer' halo.

NGC 6751 11″ SCT

NGC 6751 72mm Refractor

NGC 6755

This is a rich open cluster (over 150 stars) that is detached from the field. As you can see from the picture, there is a reasonable range of brightness.

Trumpler Classification: II 2 r

NGC 6755 11″ SCT

NGC 6755 72mm Refractor

NGC 6760

This is a loose globular cluster that will need large apertures to resolve. Note that NGC 6749 is only 1.5° away to the northwest. Can you see it (not visible in the field below)?

NGC 6760 11″ SCT

NGC 6760 72mm Refractor

NGC 6781

The brightest part of the planetary nebula is annular in shape. There is a small, faint extension on northern side. Can you see it? An UHC or an OIII filter will help.

NGC 6781 11″ SCT

NGC 6781 72mm Refractor

NGC 6804

This planetary nebula appears fan shaped, with a bright star at one end. However, with averted vision it 'grows' in size. An OIII really boosts the contrast. Can you see the fainter extensions?

NGC 6804 11″ SCT

NGC 6804 72mm Refractor

Aries

Objects	Type	Position	Size	SB*
NGC 772	Galaxy	01h59′20.1 +19 00 26	7.2×4.3	13.9

NGC 772

This is an SAb galaxy. Can you see its small, faint neighbor (NGC 770) approximately 5′ to the east?

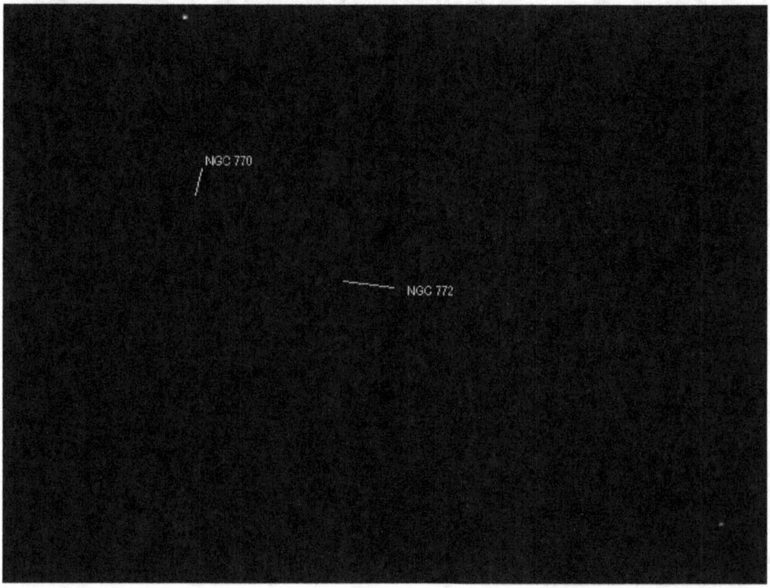

NGCs 772 and 770 11″ SCT

NGC 772 72mm Refractor

Auriga

Objects	Type	Position	Size	SB*
M 36	OC	05h36′18.0″ +34 08 00	10′	
M 37	OC	05h52′19.0″ +32 33 00	15′	
M 38	OC	05h28′43.0″ +35 51 00	15′	
NGC 1893	CN	05h22′46.0″ +33 25 00	25′	
NGC 1907	OC	05h28′05.0″ +35 19 30	5′	
NGC 1931	CN	05h31′26.0″ +34 14 42	6′	
NGC 2281	OC	06h48′18.0″ +41 05 00	25′	

M 36

This open cluster contains over 50 stars with a large range of brightness. Note the spiral-like chains of outer stars (if needed use lower magnification)!

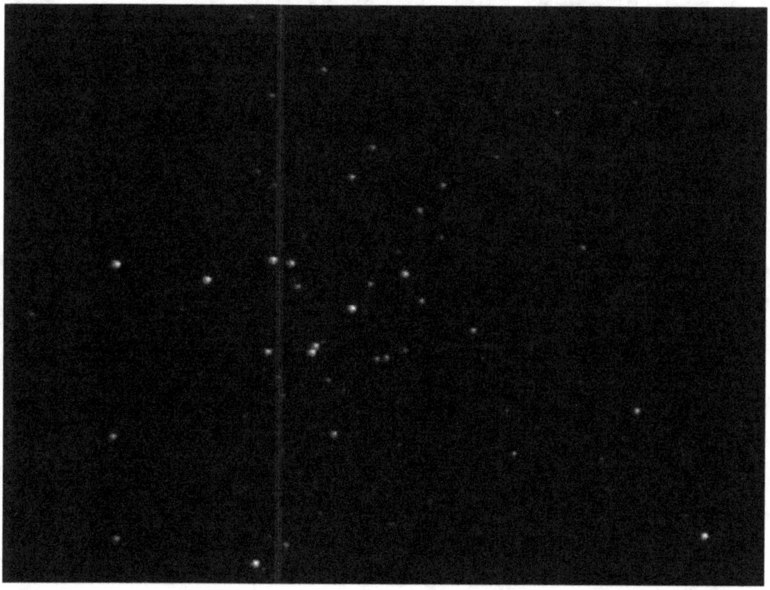

M 36 11″ SCT

M 36 72mm Refractor

Trumpler Classification: I 3 r/m

M 37

This spectacular open cluster contains almost 2,000 stars! This object responds well to high powers, revealing a multitude of stars in the core region.

Trumpler Classification: I 2 r

M 37 11" SCT

M 37 72mm Refractor

M 38

This open cluster is slightly concentrated towards the center and contains over 150 stars – although many of these are very faint. Note that NGC 1907 can be found nearby (in the same field at low power and in the refractor view below).

M 38 11″ SCT

M 38 and NGC 1907 72mm Refractor

Trumpler Classification: II 2 r

NGC 1893

This open cluster contains almost 300 stars with a large range of brightness. Can you see the nebulosity (IC 410) associated with this cluster (invisible in the pictures below)? An UHC or an OIII filter will help.

Trumpler Classification: II 3 r n

NGC 1893 11″ SCT

NGC 1893 72mm Refractor

NGC 1907

This open cluster is strongly concentrated towards the center and contains about 100 stars. It can be viewed together with M 38 at low powers – see M 38 earlier for the 72mm refractor picture of NGC 1907. There is a faint nebulosity associated with this cluster. Can you see it?

NGC 1907 11″ SCT

Trumpler Classification: I 1 m n

NGC 1931

This is a small nebulous patch that is brighter in the middle. Increasing the power will change the apparent shape of the nebulosity. An OIII or an UHC filter may help to bring out more nebulosity. How many stars can you find within the nebula?

NGC 1931 11″ SCT

NGC 1931 72mm Refractor

NGC 2281

There are over 100 stars concentrated towards the center of this apparently large open cluster. However, only 30 or so are visible in amateur telescopes.

Trumpler Classification: I 3 r/m; Older sources: I 2 p

NGC 2281 11" SCT

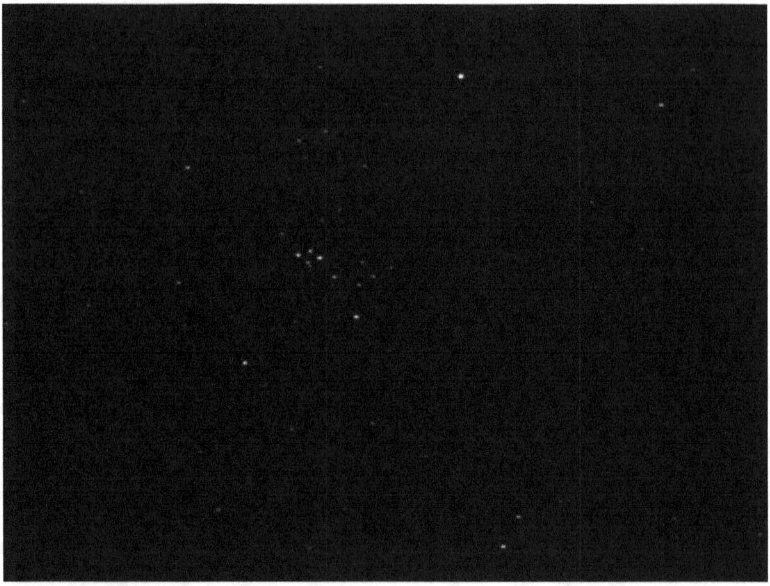

NGC 2281 72mm Refractor

Boötes

Objects	Type	Position	Size	SB*
NGC 5466	GC	14h05′27.3″+28 32 04	9.0	

NGC 5466

This is a very loose globular cluster with a low surface brightness. It very much resembles a very dim, slightly concentrated open cluster. Can you see it (invisible in the 72mm refractor)?

NGC 5466 (OCL 27) 11″ SCT

Camelopardalis

Objects	Type	Position	Size	SB*
NGC 1501	PN	04h06′59.6″ +60 55 11	52″	
NGC 1502	OC	04h07′50.0″ +62 20 00	20′	
NGC 2403	Galaxy	07h36′54.2″ +65 36 07	21.9×12.3	14.4
NGC 2655	Galaxy	08h55′37.8″ +78 13 18	4.9×4.1	13.2
Stock 23	OC	03h16′10.8″ +60 07 00	29′	

NGC 1501

This planetary nebula is sometimes called either the Oyster or Camel's Eye Nebula. The central star is a 'gettable' mag14.3. As with many planetary nebulae, this one responds well to an OIII filter. Can you detect any texture in the ring at higher magnifications?

NGC 1501 11″ SCT

NGC 1501 72mm Refractor

NGC 1502

There are over 50 stars in this open cluster with a small range of brightness. The cluster is situated at one end of the famous chain of stars, Kemble's Cascade.

Trumpler Classification: I 3 m

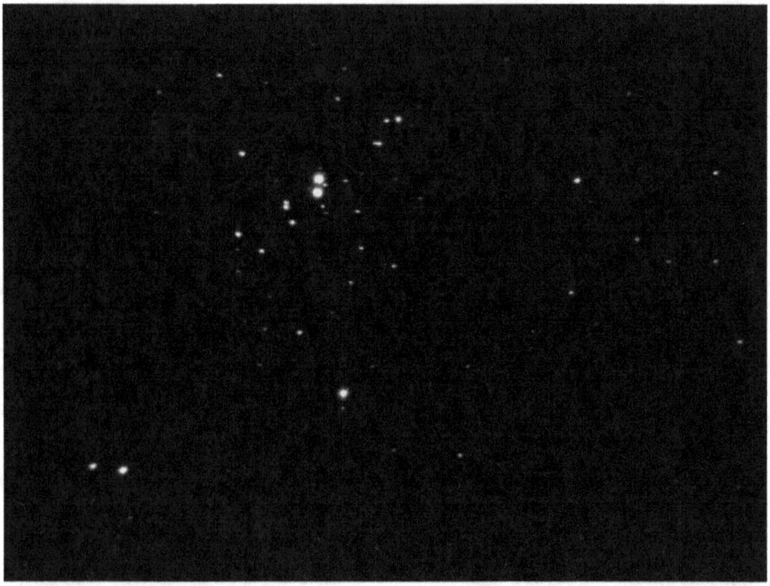

NGC 1502 11″ SCT

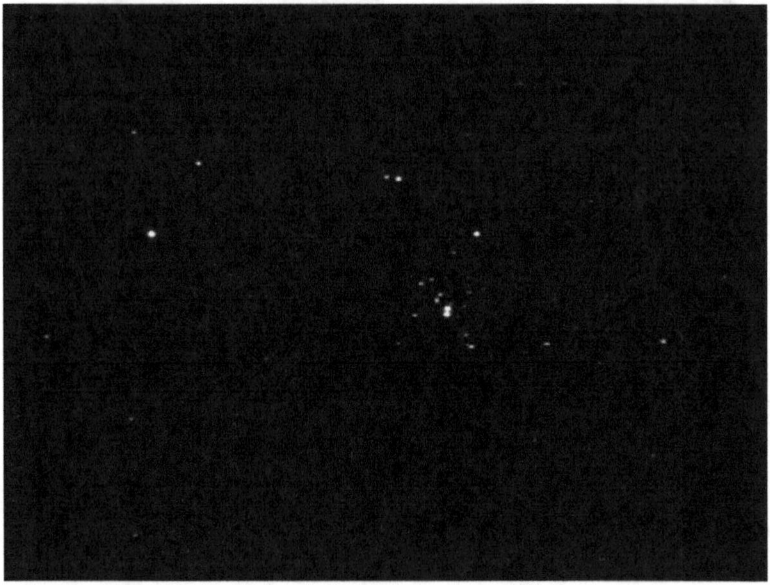

NGC 1502 72mm Refractor

NGC 2403

This is an SABcd galaxy. The core is the only partly visible in the 11″ SCT picture below. Can you see the halo and make out any structure? Note that this object was invisible in the small refractor.

NGC 2403 11″ SCT

NGC 2655

This is an SAB0 galaxy with a large extended core. The halo (if you can see it!) can appear slightly elongated.

NGC 2655 11″ SCT

NGC 2655 72mm Refractor

Stock 23

There are around 25 stars in this rather sparse open cluster, with a large range of brightness. Can you detect the faint nebulosity? This cluster is sometimes called Pazmino's Cluster.

Trumpler Classification: II/III 3 p n

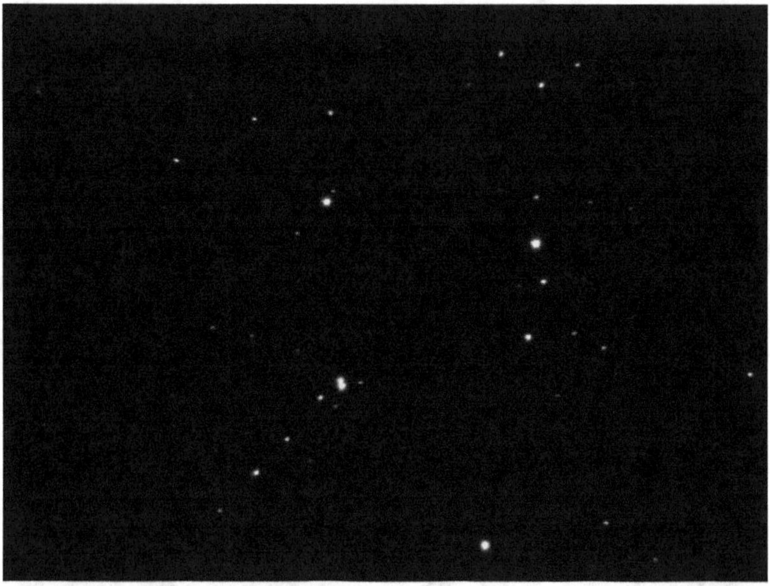

Stock 23 11″ SCT

Stock 23 72mm Refractor

Cancer

Objects	Type	Position	Size	SB*
M 44	OC	08h40′22.0″ +19 40 00	70′	
M 67	OC	08h51′24.0″ +11 49 00	25′	

M 44

This open cluster is spectacular at low powers and wide fields of view (excellent in binoculars) and is commonly known as the Beehive Cluster or Praesepe. There are over 160 stars in this cluster, although the classification says 'm,' as the majority are way beyond the reach of amateur telescopes (above mag14).

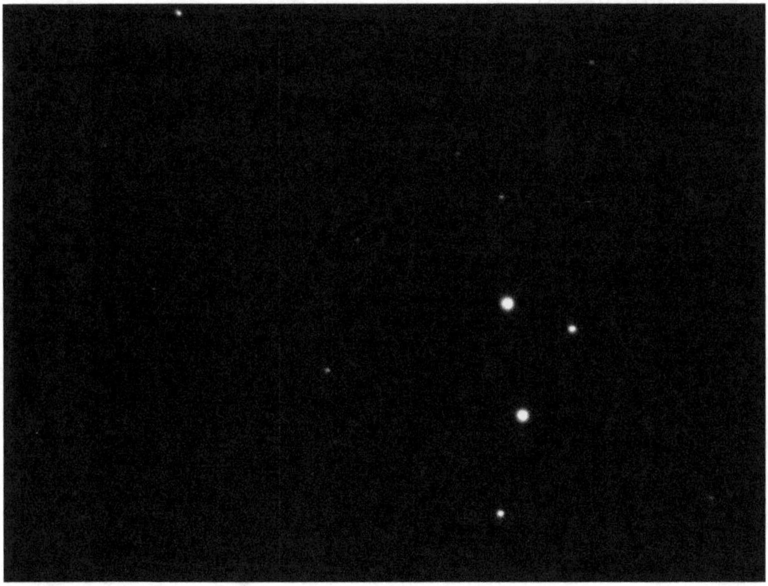

M 44 11″ SCT

M 44 72mm Refractor

Trumpler Classification: II 3 m

M 67

There is a large range of brightness in this open cluster of over 300 stars. It is more compact than M 44 (see earlier) but is just as enchanting. In contrast to its better known friend Praesepe, M 67 shows up well in telescopes, fitting nicely into medium power fields of view.

Trumpler Classification: II 3 r

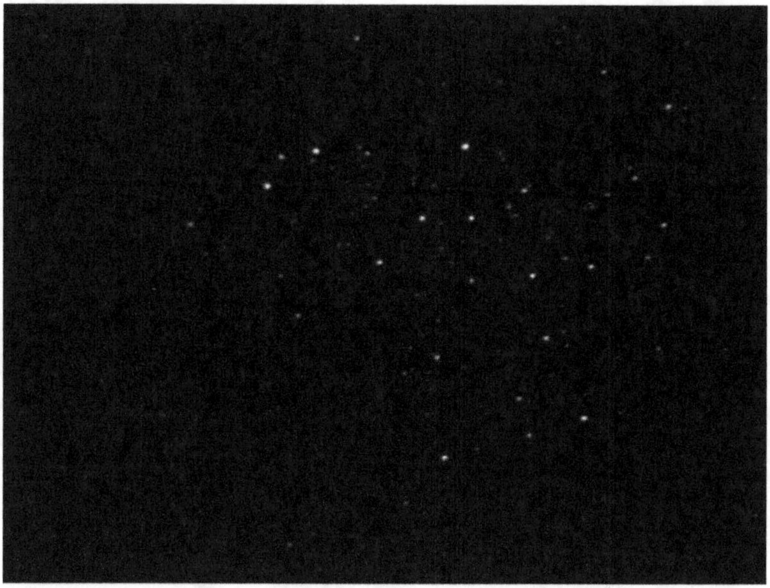

M 67 11" SCT

M 67 72mm Refractor

Canes Venatici

Objects	Type	Position	Size	SB*
M 3	GC	13h42′11.2″ +28 22 32	18.0′	
M 51	Galaxy	13h29′53.2″ +47 11 09	11.2′×6.9′	12.9
M 63	Galaxy	13h15′49.3″ +42 01 50	12.6′×7.2′	13.3
M 94	Galaxy	12h50′52.7″ +41 07 07	14.4′×12.1′	13.7
M 106	Galaxy	12h18′58.1″ +47 18 07	18.6′×7.2′	13.6
NGC 4214	Galaxy	12h15′39.0 +36 19 43	8.0′×6.6′	13.9
NGC 4244	Galaxy	12h17′29.3″ +37 48 21	16.6′×1.9′	14.0
NGC 4449	Galaxy	12h28′12.3″ +44 05 47	6.2′×4.4′	13.0
NGC 4490	Galaxy	12h30′35.8″ +41 38 26	6.3′×3.1′	12.9
NGC 4631	Galaxy	12h42′04.1″ +32 32 12	12.8′×2.4′	12.8
NGC 4656	Galaxy	12h43′54.8″ +32 09 23	10.0′×1.8′	13.5
NGC 5005	Galaxy	13h10′55.7″+37 03 26	5.8′×2.8′	12.7
NGC 5033	Galaxy	13h13′27.9″ +36 35 41	10.7′×5.0′	14.4

M 3

Note that core of this beautiful globular cluster is not so well resolved in the 72mm refractor. Can you resolve the core and make out the faint chains extending from it?

M 3 11″ SCT

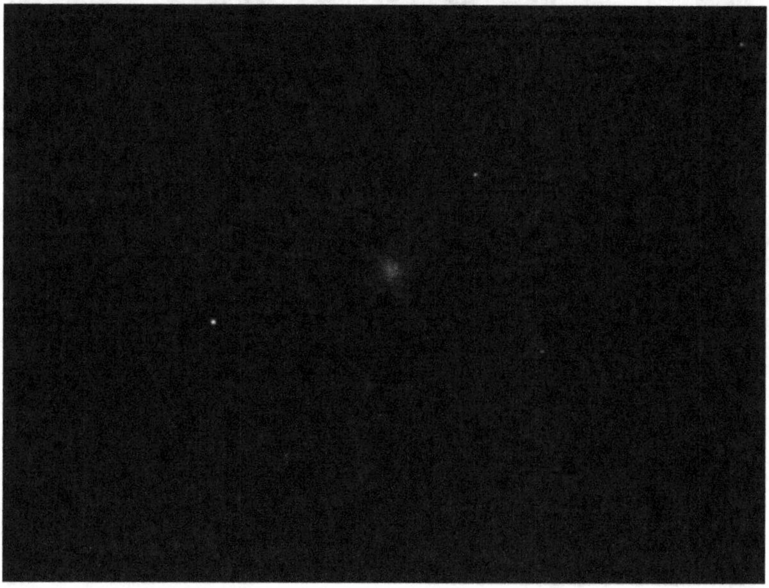

M 3 72mm Refractor

M 51

This is the famous SAbc pec commonly known as the Whirlpool Galaxy. This galaxy is interacting with nearby NGC 5195 (on the right of the 11″ SCT image below). Can you see the linking arc of material between the two and the spiral arms of M 51 – just visible in the 11″ SCT picture below?

M 51 11″ SCT

M 51 72mm Refractor

M 63

This SAbc galaxy is known as the Sunflower Galaxy. Note the bright central core. How much of the outer halo can you see? Can you see the outer spiral arms?

M 63 11″ SCT

M 63 72mm Refractor

M 94

You can clearly see the bright nucleus of this SAab galaxy. How much of the halo can you make out?

M 94 11″ SCT

M 94 72mm Refractor

M 106

This is an SABbc galaxy with a bright central core. Can you see any structure (spiral arms?) in the halo?

M 106 11" SCT

M 106 72mm Refractor

NGC 4214

This galaxy is classed IABm. Its nucleus is clearly visible in the picture below, as are a couple of knots. To what extent can you make out the structure?

NGC 4214 11″ SCT

NGC 4214 72mm Refractor

NGC 4244

This SAcd galaxy appears as a long, faint patch of nebulosity, with a very slight brightening towards one end. Can you make out the full extent of this object (approximately 17′ long)? If you can, can you see the short dark lane near the center?

NGC 4244 11″ SCT

NGC 4244 72mm Refractor

NGC 4449

Classed as an IBm galaxy, NGC 4449 has a reasonably bright central area and a stellar nucleus. Some knots are visible, and there is a star superimposed on the eastern side. To what extent can you make out the outer regions?

NGC 4449 11″ SCT

NGC 4449 72mm Refractor

NGC 4490

This galaxy is classed as an SBd pec and is sometimes called the Cocoon Galaxy. The barred shape is clearly visible. Can you make out the bright knots in the arms? Also, can you see the very faint arm linking to galaxy NGC 4485 nearby?

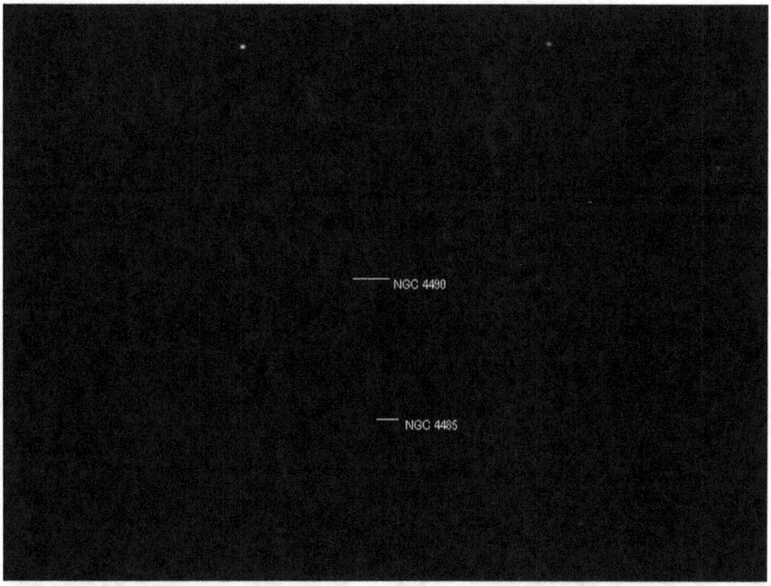

NGC 4490 11″ SCT

NGC 4490 72mm Refractor

NGC 4631

This is an edge-on SBd sp galaxy with a lot of bright and dark lanes along its length (look near the superimposed star). It is sometimes called the Whale due to its shape. It can be found in the same field of view as NGC 4656 at medium to low powers (30′ away).

NGC 4631 11″ SCT

NGC 4631 72mm Refractor

NGC 4656

This galaxy is classed as an SBm pec and has several bright patches. It is sometimes called the Hockey Stick and can be seen in the same wide field as NGC 4631. Use averted vision to spot it on the refractor image below.

NGC 4656 11" SCT

NGC 4656 72mm Refractor

NGC 5005

This is an SABbc galaxy with some dark lanes and a bright core region. It is close to NGC 5033 (approximately 1° away). Can you make out any of the dark lanes in the halo?

NGC 5005 11″ SCT

NGC 5005 72mm Refractor

NGC 5033

This Sac galaxy is close to NGC 5005 and has a bright core region with a stellar nucleus. Can you see any detail in the small halo?

NGC 5033 11″ SCT

NGC 5033 72mm Refractor

Canis Major

Objects	Type	Position	Size	SB*
M 41	OC	06h46′00.0″ −20 45 00	39′	
NGC 2345	OC	07h08′19.0″ −13 12 00	12′	
NGC 2359	EN	07h18′36.0″−13 12 00	9′×6′	
NGC 2360	OC	07h17′44.0″ −15 39 00	14′	

M 41

This is an apparently large, detached cluster with over 100 stars. It is best viewed at low powers and makes a good binocular object.

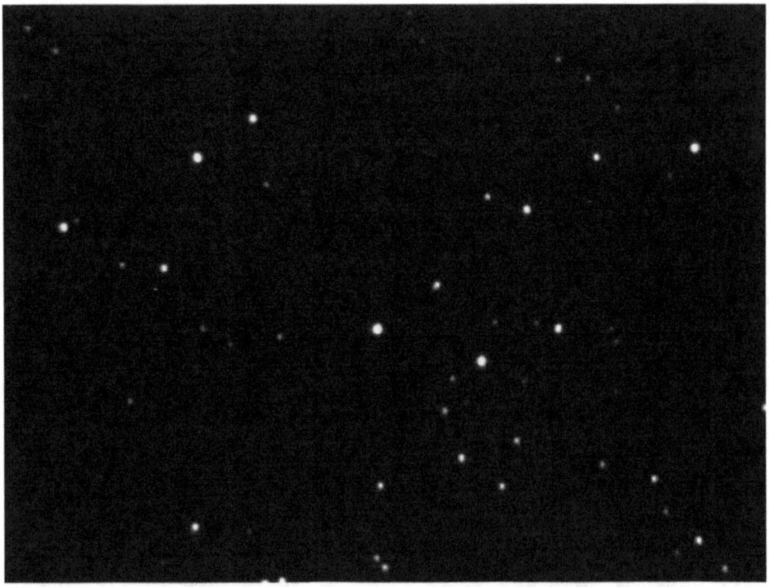

M 41 11″ SCT

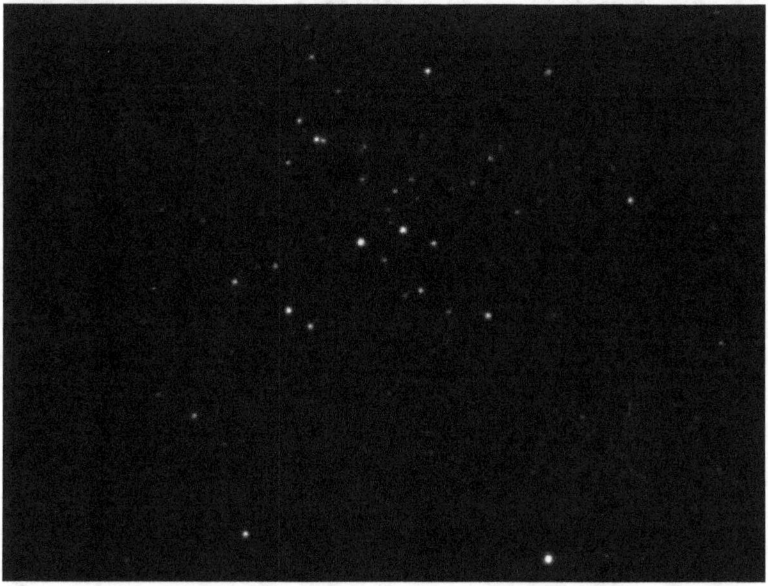

M 41 72mm Refractor

Trumpler Classification: I 3 r

NGC 2345

This young open cluster is quite rich, with a large range of brightness. It is only a short hop from NGC 2359 (see later). Note the distinctive 'V' shape of some of the brighter stars.

Trumpler Classification: II 3 r

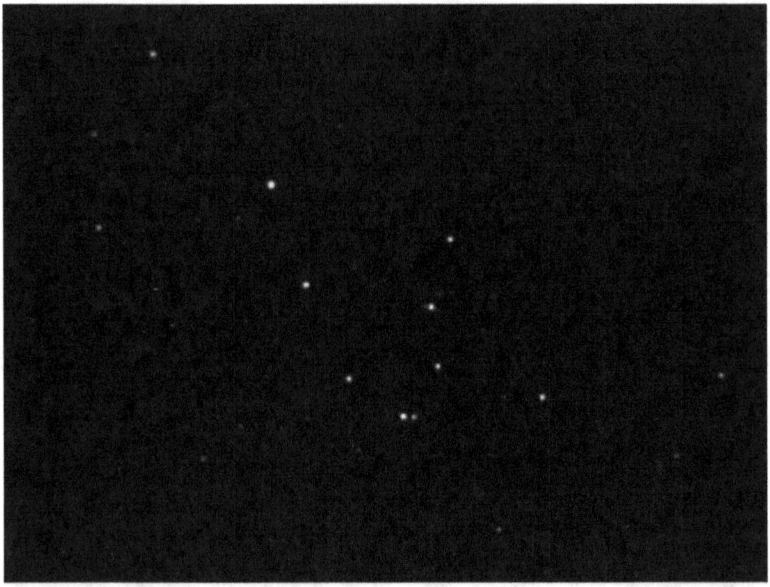

NGC 2345 11″ SCT

NGC 2345 72mm Refractor

NGC 2359

This emission nebula is noticeably brighter towards the southern end and is often called Thor's Helmet. Due to its low declination, this object is often in the low altitude haze when observed from northerly latitudes and can be very difficult to see at all. It responds well to an UHC or an OIII filter

NGC 2359 11″ SCT

NGC 2359 72mm Refractor

NGC 2360

This open cluster contains approximately 100 stars with a large range in brightness.

Trumpler Classification: I 3 r

NGC 2360 11″ SCT

NGC 2360 72mm Refractor

Capricornus

Objects	Type	Position	Size	SB*
NGC 6903	Galaxy	20h23′44.8″ −19 19 35	1.5′×1.5′	12.6

NGC 6903

This is an SAB0pec galaxy that reveals little extra detail, even at high powers. Unfortunately, due to time constraints and its low declination, the author was unable to get a decent image of this object. However, with a reasonable southerly horizon it should be visible in medium aperture telescopes.NGC 6903 11″ SCT

Cassiopeia

Objects	Type	Position	Size	SB*
Cr 26	OC	02h32′40″ +61 27 00	21′	
Cr 36	OC	03h12′00″ +63 12 00	23′	
Cr 463	OC	01h45′45″ +71 49 00	57′	
IC 289	PN	03h10′19.7″ +61 19 01	48″	
M 52	OC	23h54′51.0″ +61 36 00	16′	
M 103	OC	01h33′22.0″ +60 39 30	6′	
NGC 129	OC	00h29′58.0″ +60 13 00	12′	
NGC 185	Galaxy	00h38′57.5″ +48 20 12	8.0′×7.0′	13.6
NGC 281	CN	00h52′48.0″ +56 37 00	35′×30′	
NGC 457	OC	01h19′33.0″ +58 17 00	20′	
NGC 663	OC	01h46′17.0″ +61 13 00	15′	
NGC 7635	EN	23h20′42.0″ +61 12 00	15′×8′	
NGC 7789	OC	23h57′28.6″ +56 43 00	25′	

Cr 26

This apparently large open cluster contains approximately 50 stars. Can you make out the nebulosity associated with it?

Trumpler Classification: II 3 m n

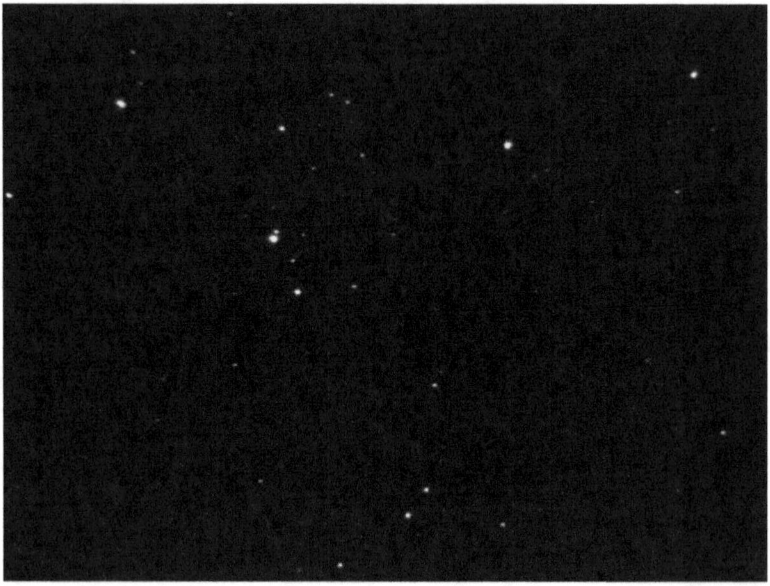

Cr 26 11″ SCT

Cr 26 72mm Refractor

Cr 36

This open cluster appears to be a larger, dimmer version of Cr 26 mentioned above and, like its friend, is best viewed at low powers.

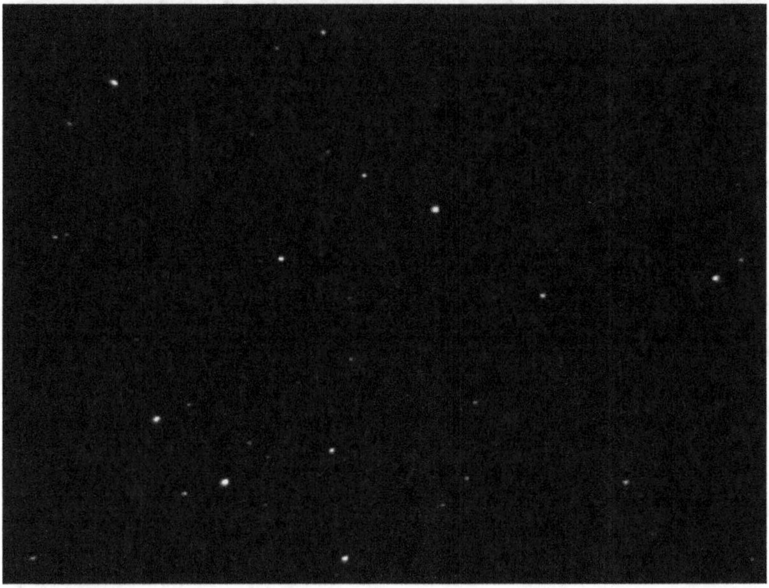

Cr 36 11″ SCT

Cr 36 72mm Refractor

Trumpler Classification: III 2 m

Cr 463

This is a very large open cluster requiring a very low magnification and a wide field of view. It can be difficult to tell where the cluster ends and the rest of the field begins!

Trumpler Classification: III 2 m

Cr 463 11″ SCT

Cr 463 72mm Refractor

IC 289

This is a very faint planetary nebula with a 10mag star close by. In the picture below it is just perceptible with averted vision. An UHC or OIII filter will help a great deal to pick this one out!

IC 289 11″ SCT

IC 289 72mm Refractor

M 52

This open cluster contains over 170 stars with a strong concentration towards the core. Note that NGC 7635, the Bubble Nebula, is less than 1° away.

Trumpler Classification: I/II 2 r

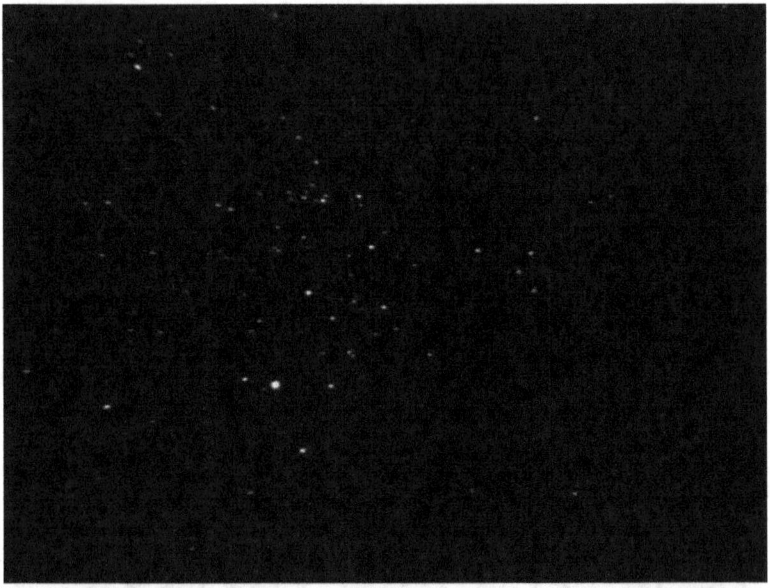

M 52 11" SCT

M 52 72mm Refractor

M 103

This open cluster contains up to 172 stars – or as little as 25, depending on which source you use! Note the Christmas tree-like pattern of the main stars.

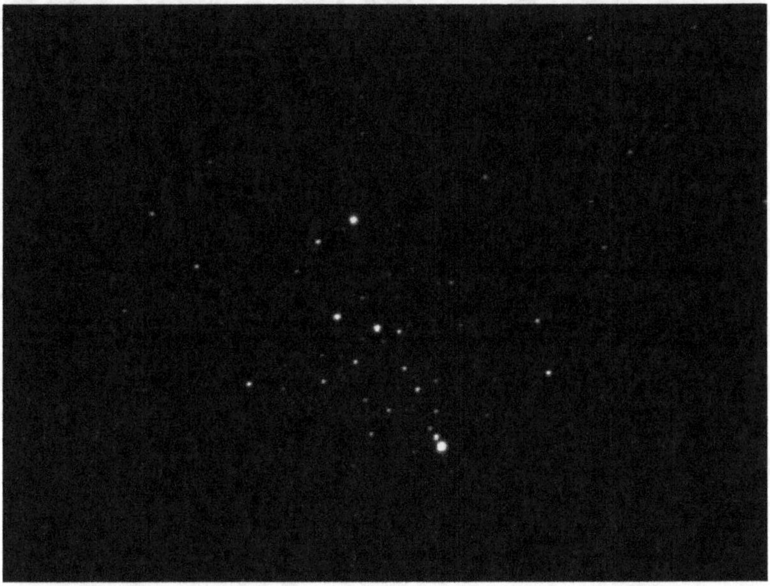

M 103 11" SCT

M 103 72mm Refractor

Trumpler Classification: II 2 m

NGC 129

This is a detached open cluster of almost 200 stars of similar brightness (although most are beyond the scope of amateur telescopes, hence the Trumpler 'm' classification).

Trumpler Classification: II/III 2 m

NGC 129 11" SCT

NGC 129 72mm Refractor

NGC 185

This rather faint galaxy is classed as an E3 pec and has some superimposed stars. Can you see the galaxy and any of the stars?

NGC 185 11" SCT

NGC 185 72mm Refractor

NGC 281

This large nebulous patch requires very low powers and an UHC or an OIII filter to capture its full extent. There is a dark intrusion from the southwest, giving it a 'Pac-man' shape. The nebulosity was invisible from the author's location. Can you see it?

NGC 281 11″ SCT

NGC 281 72mm Refractor

NGC 457

This open cluster contains over 200 stars and is sometimes referred as the Owl Cluster, or even the ET Cluster, due to the two bright 'eyes' (near the bottom of the 11″ SCT picture below) and long 'arms/wings.'

NGC 457 11″ SCT

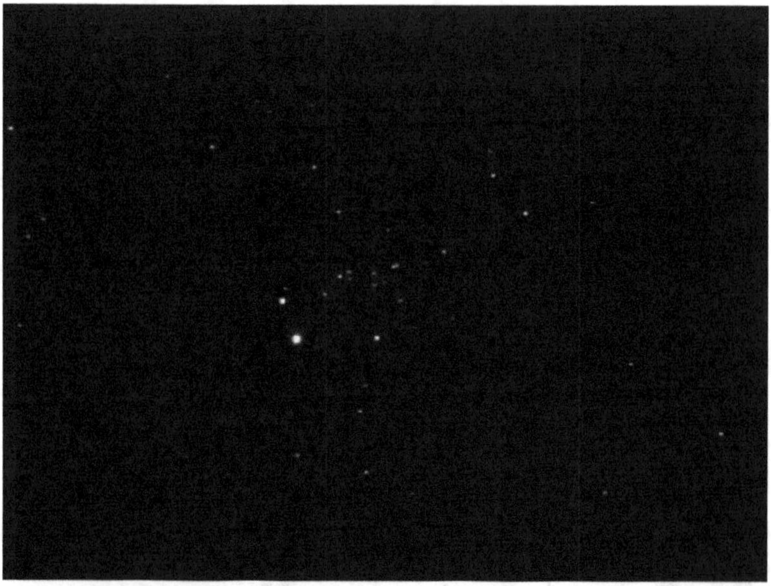

NGC 457 72mm Refractor

Trumpler Classification: II 3 r

NGC 663

There are over 100 stars in this open cluster with a large range of brightness. Note the mag9 double stars STF 151 & 152 to the west and STF 153 to the northeast of the core.

Trumpler Classification: II 3 r

NGC 663 11″ SCT

NGC 662 72mm Refractor

NGC 7635

This apparently faint emission nebula near M 52 is more commonly known as the Bubble Nebula. A part of it is visible just above center, around the bright star in the 11″ SCT picture below. Can you make out any more of the nebulosity? An UHC or OIII filter is helpful.

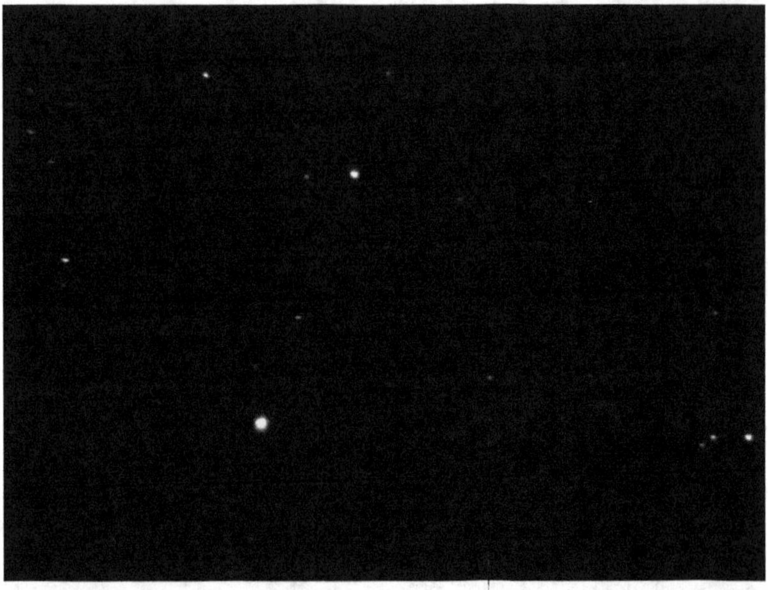

NGC 7635 11″ SCT

NGC 7635 72mm Refractor

NGC 7789

This is a very rich open cluster of almost 600 stars with quite a strong concentration towards the core.

Trumpler Classification: II 2 r

NGC 7789 11″ SCT

NGC 7789 72mm Refractor

Cepheus

Objects	Type	Position	Size	SB*
NGC 40	PN	00h13′00.9″ +72 31 19	74″	
NGC 6939	OC	20h31′30.0″ +60 40 00	10′	
NGC 6946	Galaxy	20h34′54.9″ +60 09 23	11.5′×9.8′	13.8
NGC 7023	OC+RN	21h01′37.0″ +68 10 00	10′×8′	
NGC 7129	OC+RN	21h42′00.0″ +66 05 00	8′	
NGC 7160	OC	21h53′40.0″ +62 36 12	5′	
NGC 7235	OC	22h12′24.0″ +57 16 24	6′	

NGC 40

This is a reasonably bright, large, and slightly elongated planetary nebula. An UHC filter will increase the contrast.

NGC 40 11″ SCT

NGC 40 72mm Refractor

NGC 6939

There are approximately 300 stars in this open cluster, with moderate concentration towards the center. Note that galaxy NGC 6946 is approximately 40′ away.

Trumpler Classification: II 1 r

NGC 6939 11″ SCT

NGC 72mm Refractor

NGC 6946

This is an SABcd galaxy, close to NGC 6939 above, with a bright center and some superimposed stars. This spiral galaxy is presented to us 'face-on' and has a very low surface brightness. Can you see the spiral arms?

NGC 6946 11" SCT

NGC 6946 72mm Refractor

NGC 7023

This reflection nebula contains open cluster Cr 427. Notice the 'gap' around the central star and other dark 'arcs' of dust? This nebula does not respond particularly to an UHC filter, although a Neodymium filter adds a little to the contrast.

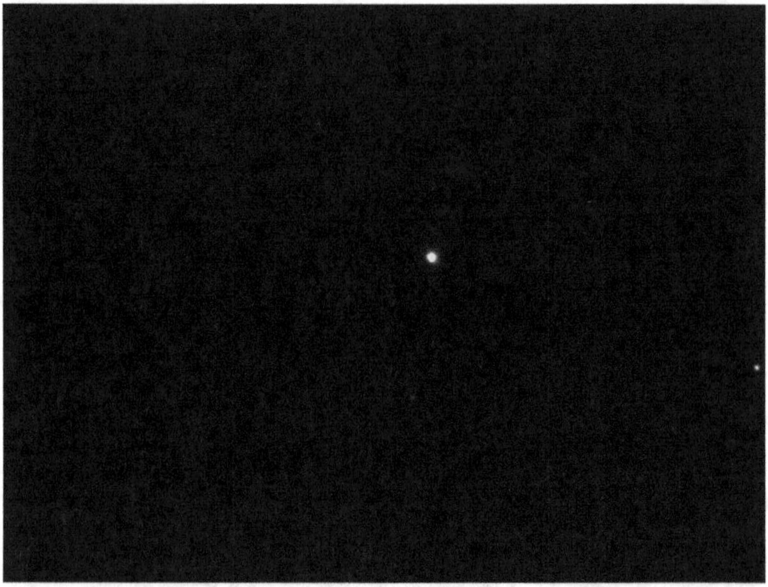

NGC 7023 11″ SCT

NGC 7023 72mm Refractor

NGC 7129

This is a small cluster of ten stars surrounded by a faint and irregularly shaped reflection nebula. Can you see all ten stars? Can you see any nebulosity (some nebulosity is just visible in the 11″ SCT picture below around the stars)?

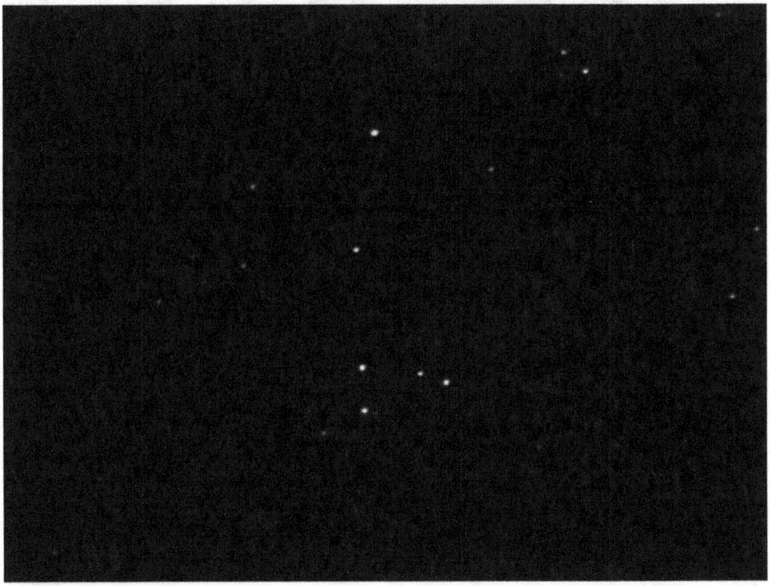

NGC 7129 11″ SCT

NGC 7129 72mm Refractor

Trumpler Classification: IV 2 p n

NGC 7160

There are approximately 50 stars in this open cluster, which is well detached from the field. There is a large range of brightness. The challenge is to see how many you can pick out!

Trumpler Classification: I 3 p

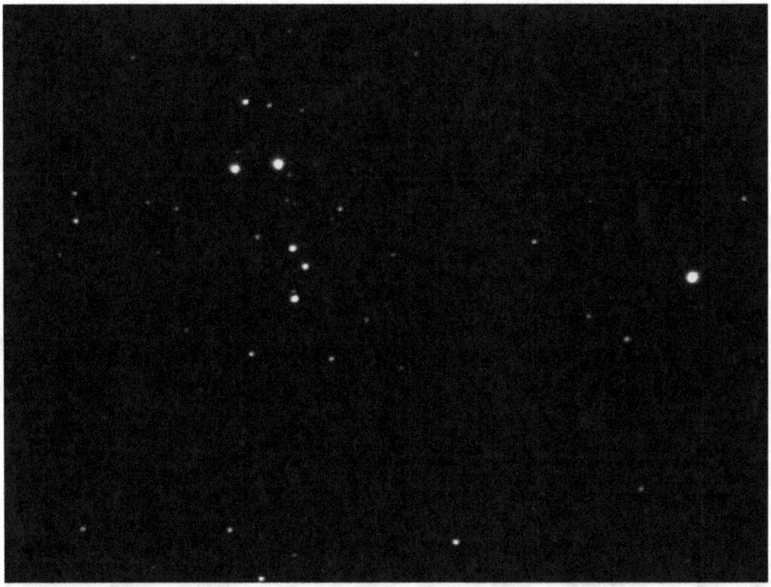

NGC 7160 11″ SCT

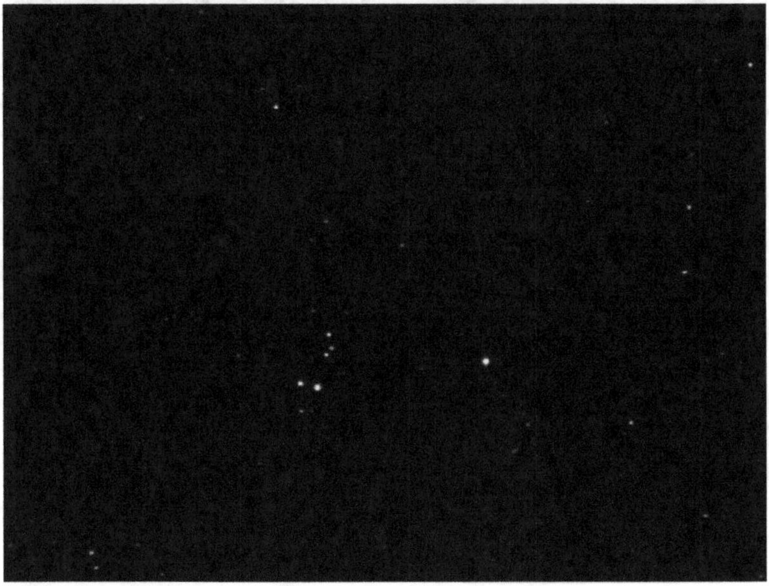

NGC 7160 72mm Refractor

NGC 7235

There are almost 100 stars in a wide range of brightness's in this cluster. Can you see the planetary nebula Minkowski 2–51(or PK103+0.1, RA=22h16′01.8″, Dec=57 28 33) only 30′ away?

Trumpler Classification: II 3 m

NGC 7235 11″ SCT

NGC 7235 72mm Refractor

Cetus

Objects	Type	Position	Size	SB*
M 77	Galaxy	02h42′40.9″ −00 00 43	7.1′×6′	12.8
NGC 246	PN	00h47′03.6″ −11 52 20	245″	
NGC 936	Galaxy	02h27′37.0″ −01 09 19	4.7′×4.1′	13.2

M 77

This SAb galaxy has a bright core with a superimposed star just 1.2′ from the center. Large apertures (and high powers) are required to see a hint of the spiral arms. Can you see them?

M 77 11″ SCT

M 77 72mm Refractor

NGC 246

This is a large, low surface brightness planetary with several stars superimposed. Can you see this object? It is visible only with averted vision in the 11″ SCT picture below. An UHC or an OIII filter will help to bring out extra detail in the nebula, particularly the latter.

NGC 246 11″ SCT

NGC 936

This is an SB0 galaxy with a short bar and an overall oval-shaped halo. Can you make out any structure?

NGC 936 11″ SCT

NGC 936 72mm Refractor

Coma Berenices

Objects	Type	Position	Size	SB*
Abell 1656	Gx Cluster	12h59′48.0″ +27 58 00	220′	
Mel 111	OC	12h25′06.0″ +26 07 00	300′	
M 53	GC	13h12′55.3″ +18 10 09	13′	
M 64	Galaxy	12h56′43.9″ +21 40 54	10.0′×5.4′	12.7
M 85	Galaxy	12h25′24.1″ +18 11 18	7.1′×5.5′	13.0
M 88	Galaxy	12h31′59.5″ +14 25 15	6.9′×3.7′	13.0
M 91	Galaxy	12h35′26.2″ +14 29 45	5.4′×4.3′	13.4
M 98	Galaxy	12h13′48.8″ +14 53 52	9.8′×2.8′	13.6
M 99	Galaxy	12h18′48.1″ +14 25 16	5.4′×4.7′	13.2
M 100	Galaxy	12h22′55.9″ +15 49 12	7.4′×6.3′	13.4
NGC 4274	Galaxy	12h19′50.3″ +29 36 52	6.8′×2.5′	13.3
NGC 4293	Galaxy	12h21′13.2″ +18 23 03	5.6′×2.6′	13.1
NGC 4414	Galaxy	12h26′27.3″ +31 13 17	4.4′×3.0′	12.8
NGC 4494	Galaxy	12h31′24.3″ +25 46 30	4.8′×3.5′	12.9
NGC 4559	Galaxy	12h35′57.7 +27 57 52	10.7′×4.4′	14.0
NGC 4565	Galaxy	12h36′21.5″ +25 59 07	15.8′×2.1′	13.2
NGC 4725	Galaxy	12h50′26.7″ +25 30 12	10.7′×7.6′	14.0

Abell 1656

This is best known as the Coma Cluster, with over 100 galaxies within a 220′ region. NGCs 4874 and 4889 are found near the central region. Note the several faint fuzzies visible in the 11″ SCT picture below. How many galaxies can you see?

Abell 1656 11″ SCT

Mel 111

This is a very large open cluster, visible to the naked eye, even in suburban locations, as a large fuzzy patch. There are over 200 stars to be found. Very low powers (binoculars?) are best for this object. It is included it here even though there is no image below, as neither of the author's two optical instruments can truly do Mel 111 justice – it's simply too large!

Trumpler Classification: III 3 r

M 53

This is a beautiful globular cluster with a few, short bright chains extending from the core. How much of the halo of stars around the core can you resolve?

M 53 11″ SCT

M 53 72mm Refractor

M 64

This famous SAab a galaxy is better known as the Blackeye Galaxy. From the picture below, the reasons for this are obvious. How much of the halo can you make out? Can you see either of the spiral arms?

M 64 11″ SCT

M 64 72mm Refractor

M 85

This is a SA0 pec galaxy with NGC 4394 in the same field of view. Note also the mag10 star at the southeastern end of the galaxy and a mag13 star superimposed near the northern end.

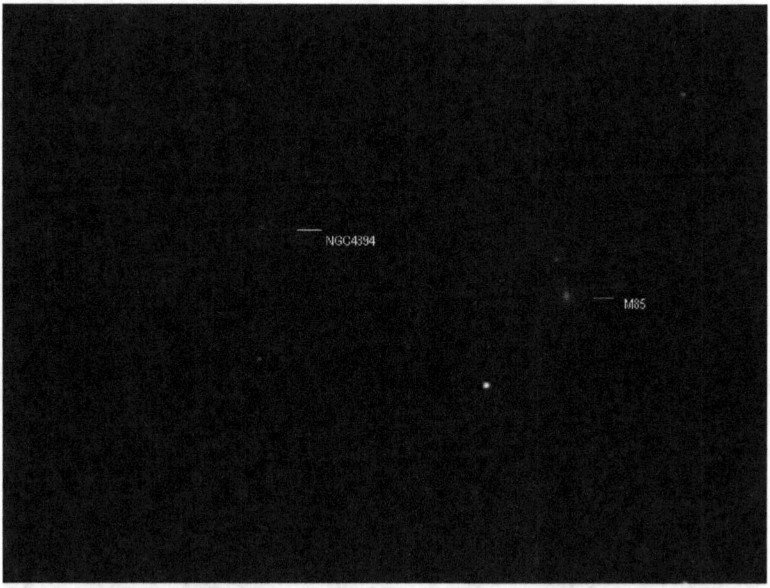

M 85 11″ SCT

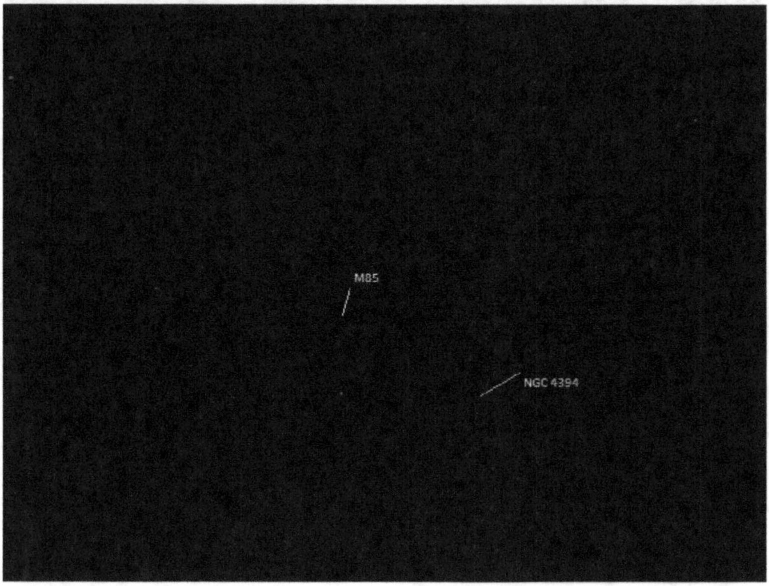

M 85 72mm Refractor

M 88

This is a SAb galaxy presented to us almost edge-on. It has an almost stellar core. Can you make out any structure in the halo? Note the superimposed mag11 double star.

M 88 11″ SCT

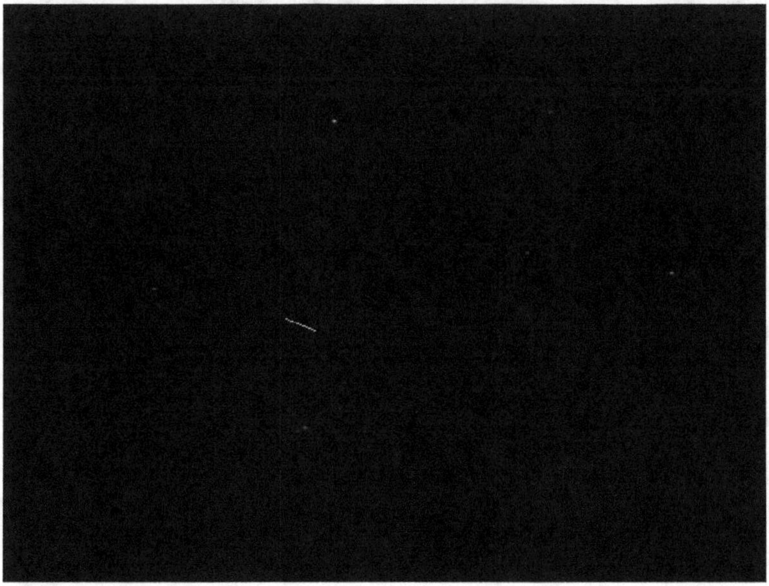

M 88 72mm Refractor

M 91

This is a SBb galaxy with a small, slightly elongated halo.

M 91 11″ SCT

M 91 72mm Refractor

M 98

This is a SABab galaxy with a large, bright core and stellar nucleus. Can you make out any mottling or other detail in the halo?

M 98 11" SCT

M 98 72mm Refractor

M 99

This is a SAc galaxy presented to us face-on and has a stellar nucleus. Can you make out both of the spiral arms (just visible in the 11" SCT picture below)?

M 99 11″ SCT

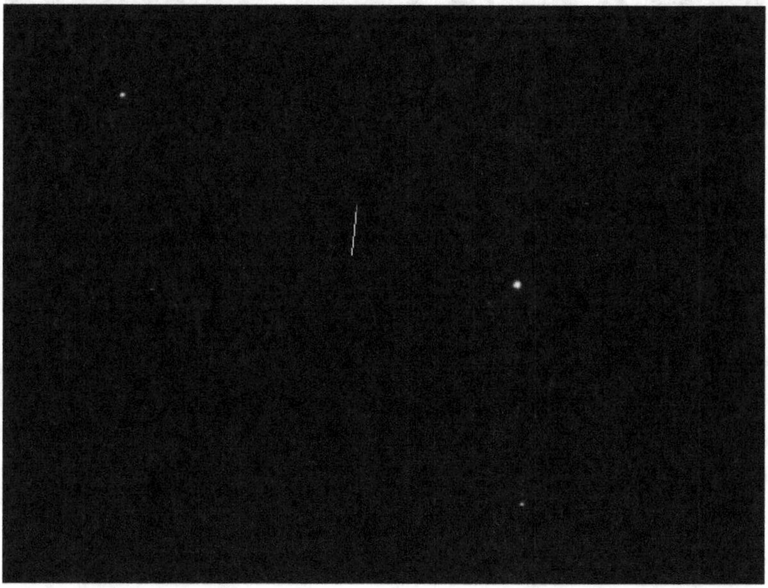

M 99 72mm Refractor

M 100

This SABbc galaxy has a bright core. Spiral structure is visible in the halo in larger apertures (just visible in the 11″ SCT picture below). Can you see any structure?

M 100 11″ SCT

M 100 72mm Refractor

NGC 4274

This is an SBab galaxy, presented to us almost face-on. You can clearly see the bright core and the spiral arms are just visible in the 11″ SCT image below. Can you see them?

NGC 4274 11″ SCT

NGC 4274 72mm Refractor

NGC 4293

This Sbab galaxy has a reasonably bright nucleus and an elongated halo. Can you make out any structure?

NGC 4293 11″ SCT

NGC 4293 72mm Refractor

NGC 4414

This is an Sac galaxy and has a bright core region surrounded by a small, faint halo. The halo appears elongated in the 11″ SCT image below.

NGC 4414 11″ SCT

NGC 4414 72mm Refractor

NGC 4494

This E1-2 galaxy has a bright center and appears slightly elongated. Otherwise, there appears to be little extra detail that can be gleaned from this object.

NGC 4494 11″ SCT

NGC 4494 72mm Refractor

NGC 4559

This is an SABcd galaxy with a faint, non-stellar core and faint halo. If you can see it, can you make out any detail in the halo? This object is just visible in the images below using averted vision.

NGC 4559 11″ SCT

NGC 4559 72mm Refractor

NGC 4565

The bisecting dust lane is very obvious in this SAb (?) galaxy. Can you see its full length (Hint: use averted vision)? Can you see the mottling to the southeast of the core, along the dust lane?

NGC 4565 11″ SCT

NGC 4565 72mm Refractor

NGC 4725

This SABab pec galaxy has a bright core and a small bar. There are some very faint, superimposed stars. Can you make out any structure in the halo (spiral arms)? Note that, in the 11" SCT image below, this detail is barely perceptible. Use averted vision.

NGC 4725 11″ SCT

NGC 4725 72mm Refractor

Corona Borealis

Objects	Type	Position	Size	SB
NGC 6085	Galaxy	16h12′35.4″ +29 21 54	1.5′×1.2′	13.5
NGC 6086	Galaxy	16h12′35.4″ +29 29 02	1.7′×1.2′	13.5

NGC 6085 and 6086

These two galaxies appear small and in the same medium power field of view. They are classed Sa and E, respectively. NGC 6085 has a brighter nucleus than its friend, whereas NGC 6086 has a foreground star on its northwestern edge.

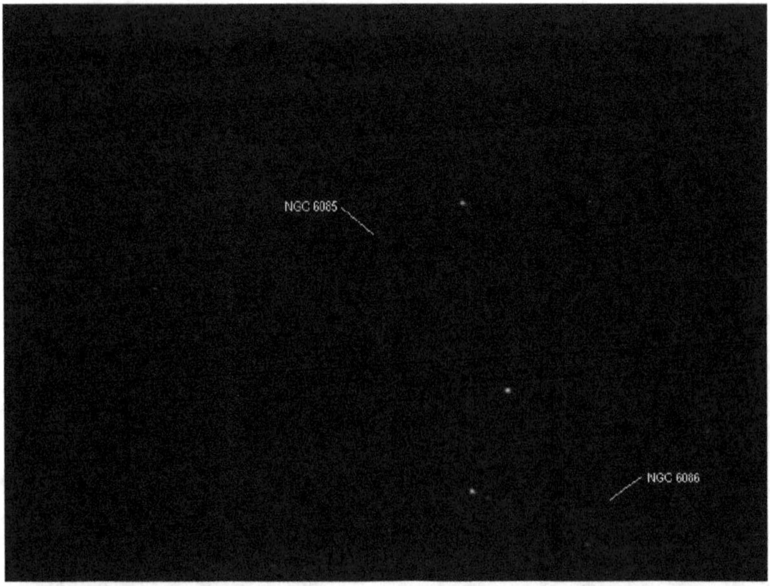

NGCs 6085 and 6086 11″ SCT

Corvus

Objects	Type	Position	Size	SB*
NGC 4038	Galaxy	12h01′59.2″ −18 52 05	3.4′×1.7′	12.0
NGC 4039	Galaxy	12h01′54.3″ −18 53 31	3.2′×2.1′	12.2
NGC 4361	PN	12h24′30.8″ −18 47 02	126″	10.2

NGCs 4038 and 4039

NGC 4038 is classed SBm pec and is commonly known as the Antennae as it is colliding with SA galaxy, NGC 4039. This is a difficult object in northerly latitudes due to its low altitude. How much of the knotted structure can you make out?

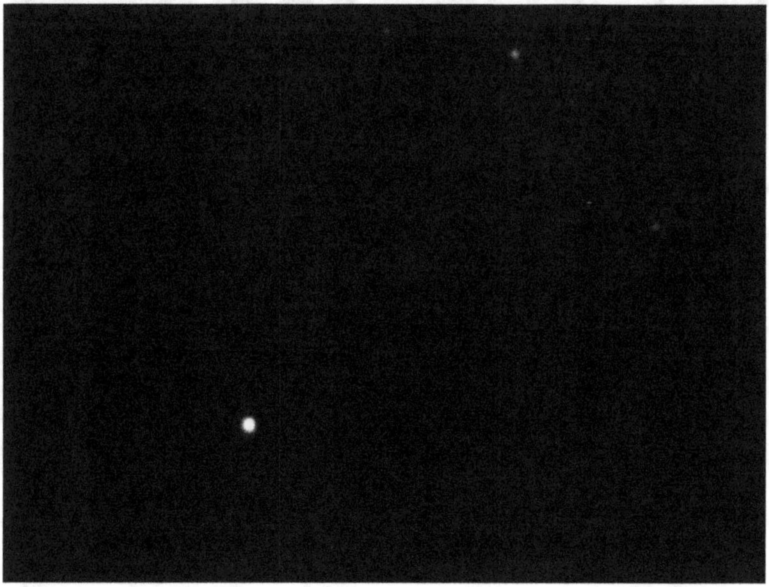

NGCs 4048 and 4039 11″ SCT

NGC 4361

This planetary nebula seems to have a 'mottled' appearance at higher powers due to the brighter central area. The central star is 13th magnitude and quite easy to see. How much of the outer halo can you see? Using an UHC filter adds to the contrast.

NGC 4361 11″ SCT

NGC 4361 72mm Refractor

Crater

Objects	Type	Position	Size	SB*
NGC 3636	Galaxy	11h20′25.1″–10 16 55	1.3′×1.3′	13.0
NGC 3637	Galaxy	11h20′39.5″ –10 15 25	1.9′×1.6′	13.8
NGC 3672	Galaxy	11h25′02.5″ –09 47 40	4.4′×1.9′	13.5
NGC 3892	Galaxy	11h48′01.2″ –10 57 45	3.0′×2.2′	13.4

NGC 3636 and 3637

This E0 and SB pair of galaxies can be found either side of a 6.5mag star. Note that they are barely visible in the 72mm refractor view below.

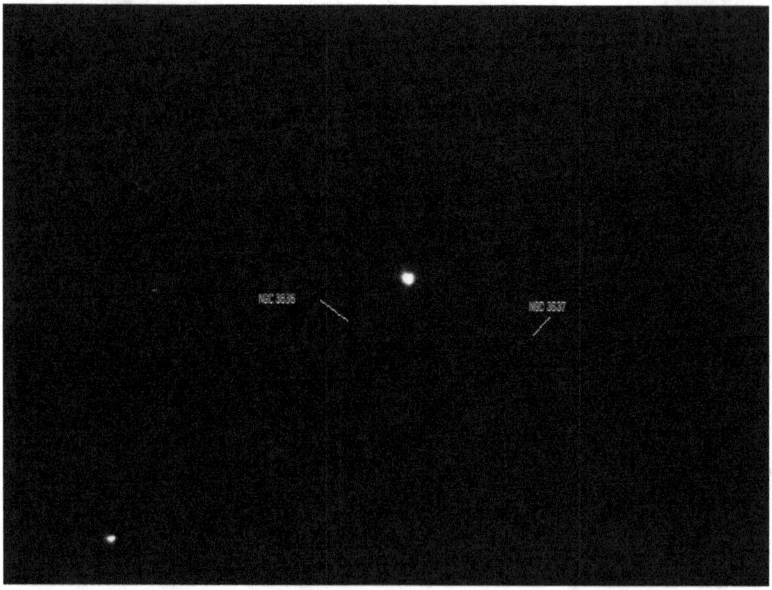

NGCs 3636 and 3637 11″ SCT

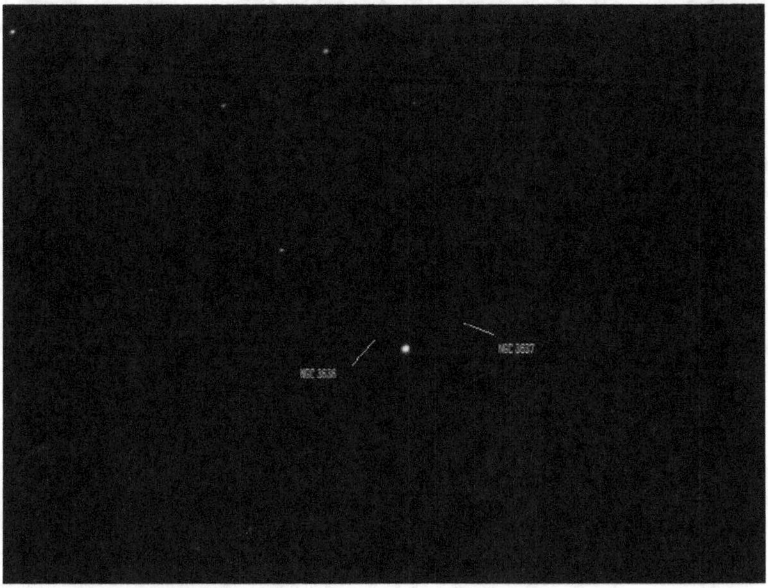

NGCs 3636 and 3637 72mm Refractor

NGC 3672

This apparently faint Sac galaxy has a 10th mag star only 4′ away and is presented to us at a slight angle, hence its elongated appearance. Note that it is barely perceptible in the 72mm refractor view below.

NGC 3672 11″ SCT

NGC 3672 72mm Refractor

NGC 3892

This galaxy is classed as a SB0 galaxy. How much of the halo can you make out? Can you see any structure?

NGC 3892 11″ SCT

NGC 3892 72mm Refractor

Cygnus

Objects	Type	Position	Size	SB*
M 29	OC	20h24′06.0″ +38 30 00	10′	
M 39	OC	21h31′52.0″ +48 26 00	31′	
NGC 6819	OC	19h41′18.0″ +40 11 00	5′	
NGC 6826	PN	19h44′48.2″ +50 31 31	36″	
NGC 6888	EN	20h12′00.0″ +38 21 00	18′×13′	
NGC 6910	OC	20h23′12.0″ +40 47 00	10′	
NGC 6960	SR	20h45′42.0″ +30 43 00	70′×6′	
NGC 6992/6995	SR	20h56′24.0″ +31 43 00	60′×8′	
		20h57′06.0″ 31 13 00	12′×12′	
NGC 7000	EN	20h58′48.0″ +44 20 00	120′×100′	
NGC 7008	PN	21h00′33.1″ +54 32 32	86″	
NGC 7026	PN	21h06′18.5″ +47 51 08	45″	
NGC 7027	PN	21h07′01.8″ +42 14 07	55″	
NGC 7063	OC	21h24′30.0″ +36 30 00	9′	

M 29

This detached cluster contains nearly 100 stars of varying brightness. Most of these stars are at the limit of visibility and imbedded in faint nebulosity. Can you make out this nebulosity?

Trumpler Classification: II 3 m n

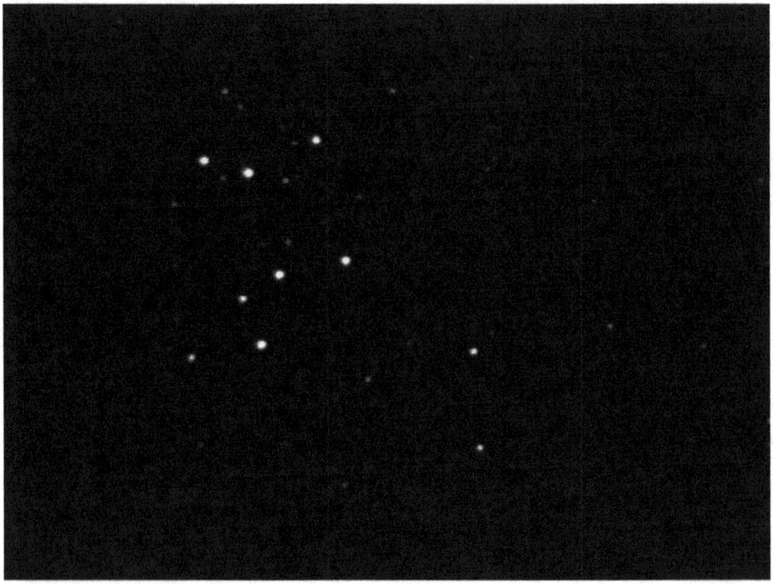

M 29 11″ SCT

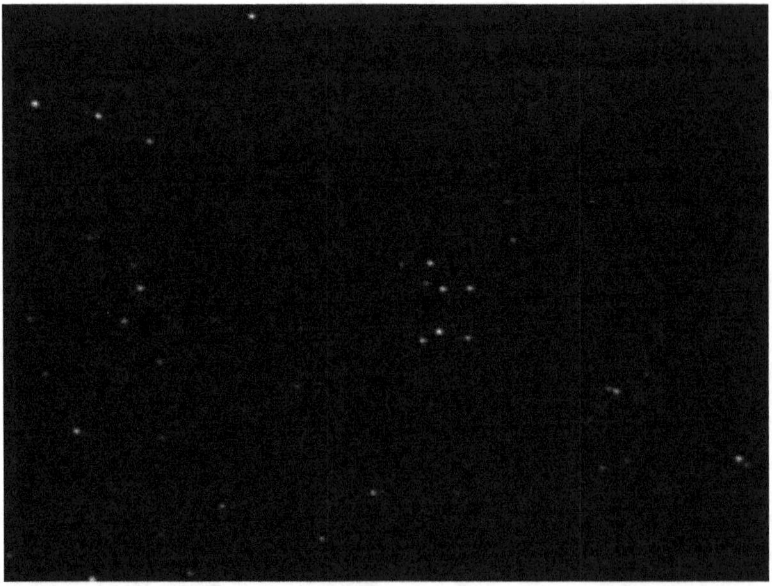

M 29 72mm Refractor

M 39

This open cluster contains about 30 stars detached from a Milky Way background, forming a large triangular shape. It is best viewed using low powers and wide fields of view (binoculars?).

Trumpler Classification: III 2 m/p

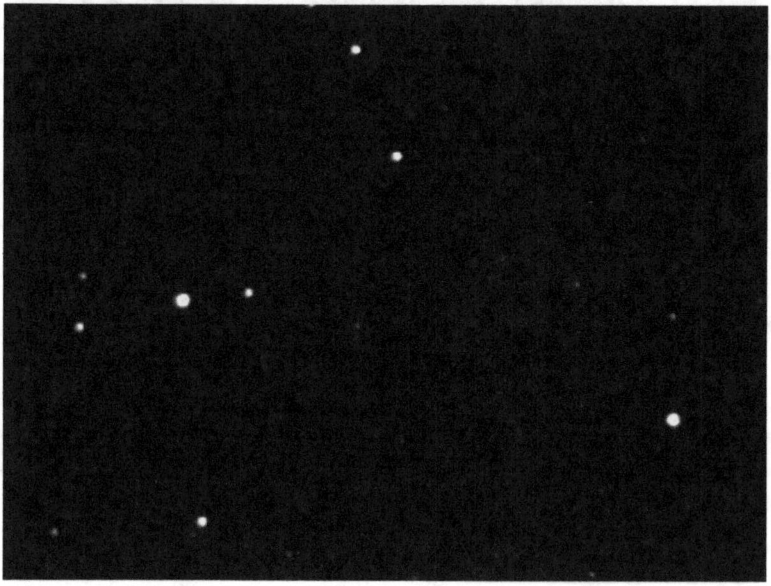

M 39 11″ SCT

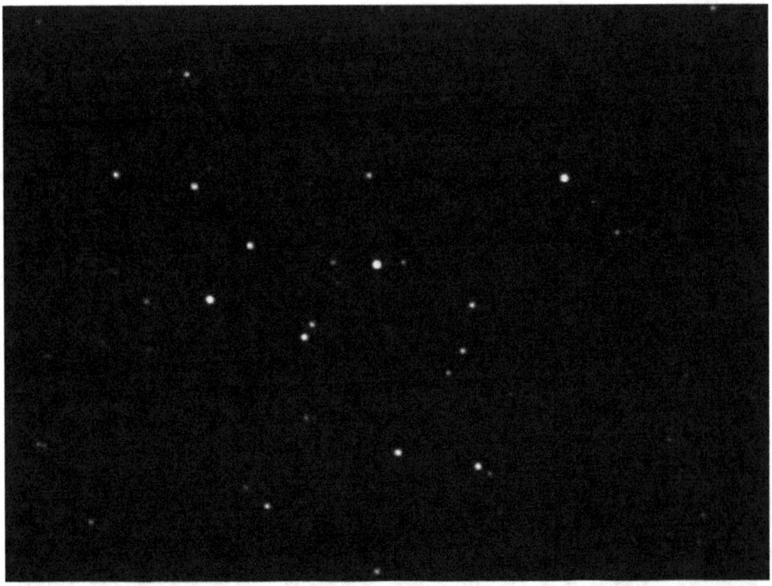

M 39 72mm Refractor

NGC 6819

This open cluster is strongly concentrated towards the center and contains almost 1,000 stars – most of which are beyond the reach of most amateur telescopes!

Trumpler Classification: I 1 r

NGC 6819 11″ SCT

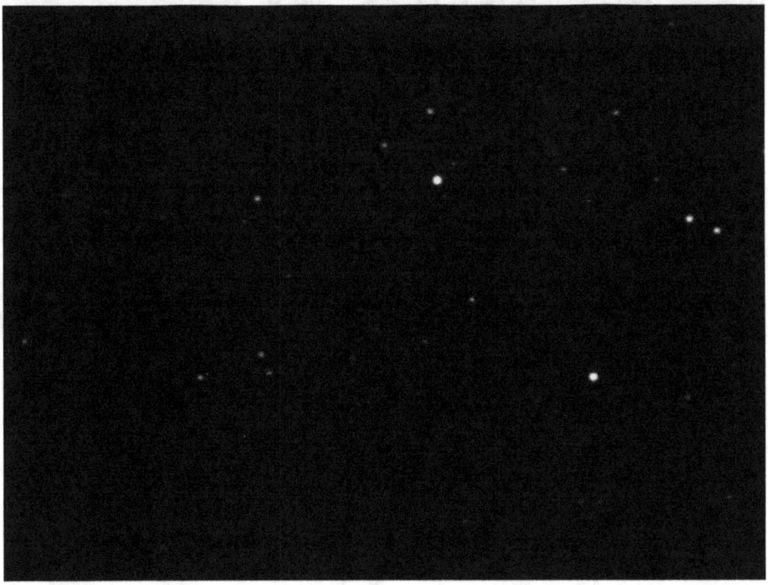

NGC 6819 72mm Refractor

NGC 6826

This light green planetary nebula is more commonly known as the Blinking Planetary – switch between direct and averted vision to see the effect – and has a bright central star. The nebula appears significantly larger with averted vision. Note that the 'blinking' effect is lost in larger apertures, but more inner detail is visible.

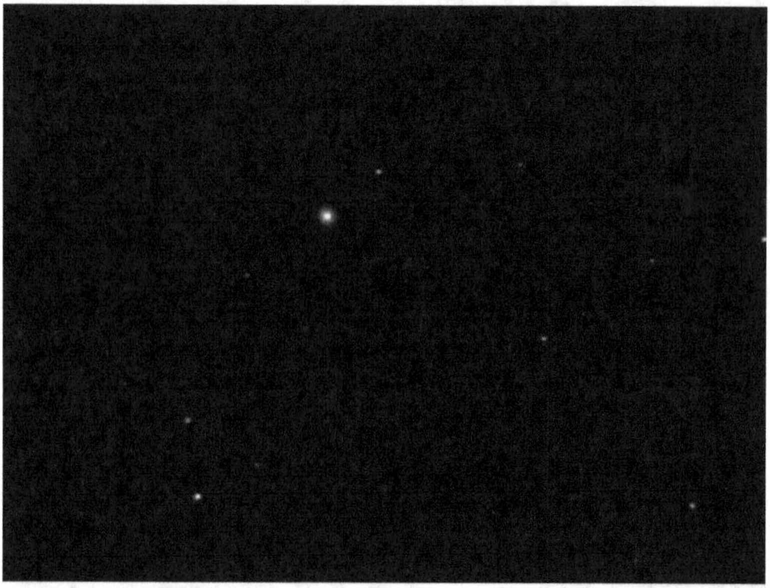

NGC 6826 11″ SCT

NGC 6826 72mm Refractor

NGC 6888

This emission nebula comprises of a broken ring and is often called the Crescent Nebula. It is just perceptible in the pictures below. This object is best observed using UHC or OIII filters. How much of it can you see?

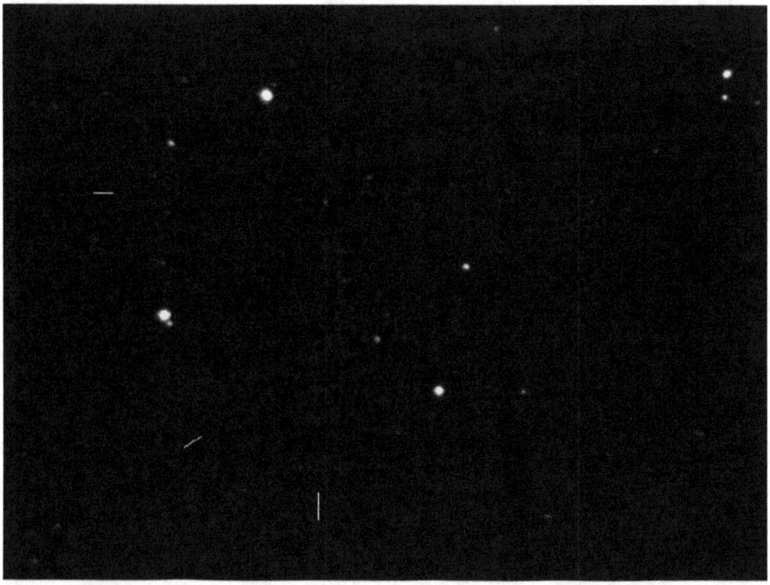

NGC 6888 11″ SCT

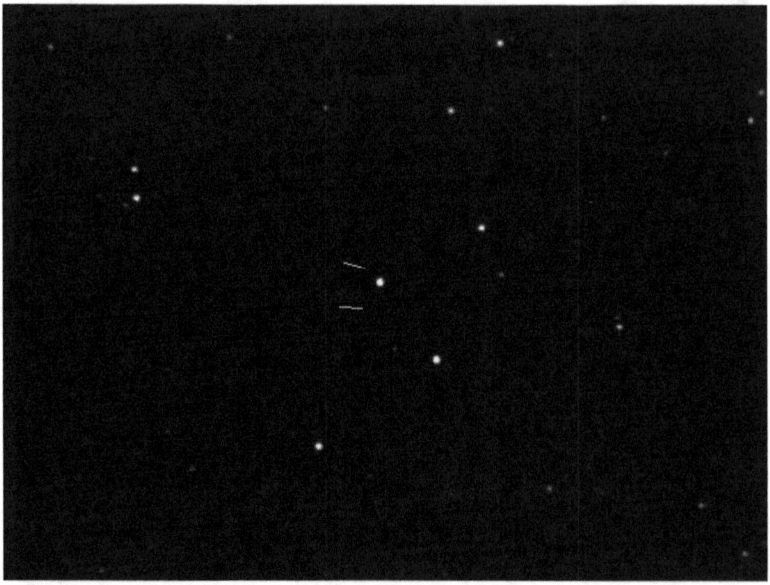

NGC 6888 72mm Refractor

NGC 6910

This open cluster contains approximately 60 stars (most of which are very dim), concentrated towards the center, and is also surrounded by nebulosity (IC 1318). Can you detect the nebulosity?

Trumpler Classification: I 3 m n

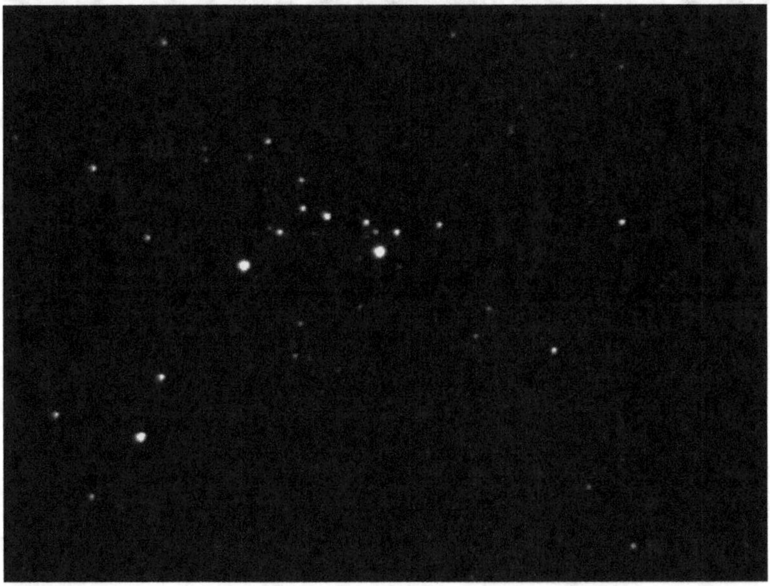

NGC 6910 11″ SCT

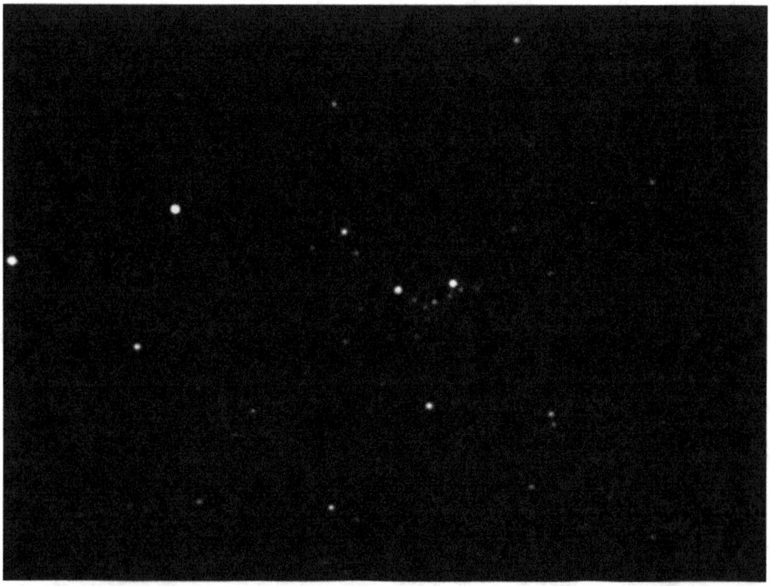

NGC 6910 72mm Refractor

NGC 6960

This forms the western part of the famous Veil Nebula (a supernova remnant) and is also the brightest part (see the bright star 52 Cygnus near the center in the picture below). It is also sometimes known as the Witch's Broom. This object should be viewed together with the rest of the structure (NGCs 6974, 6979, 6992, and 6995) using low powers. As this is also an emission nebula, an OIII filter works particularly well and will aid detection in smaller apertures. High powers should bring out some remarkable detail.

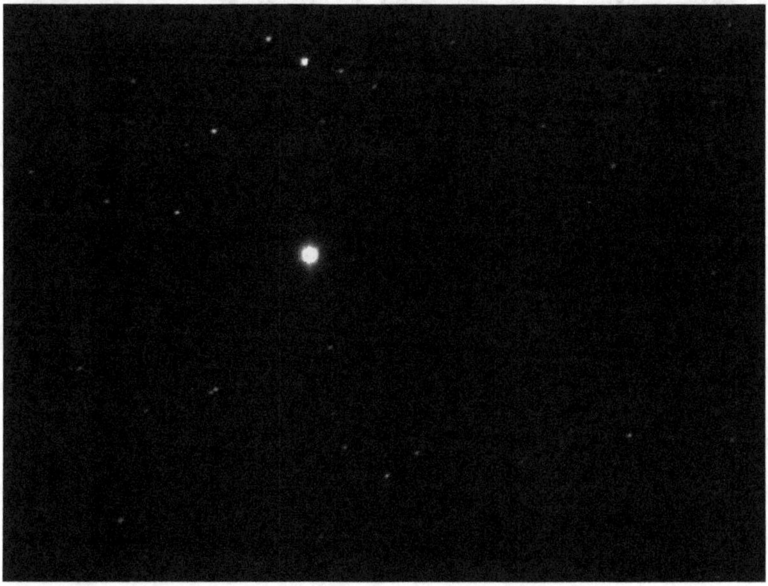

NGC 6960 11″ SCT

NGCs 6992 and 6995

These objects form the other sections of the famous Veil Nebula (see NGC 6960). This entire complex is also known by other common names, including Cirrus, Filamentary, Network, and Cygnus Loop. In the 11″ SCT picture below, NGC 6992 is just visible as an extremely faint nebulosity (use averted vision) running through the center of the field from top to bottom. As mentioned with NGC 6960, an OIII filter will greatly increase the likely-hood of detecting these sections of the nebula.

NGC 6992 11″ SCT

NGC 6992 72mm Refractor

NGC 7000

Due to its distinctive shape, this emission nebula is more commonly known as the North American Nebula and is best viewed using low powers and wide fields of view, i.e., a small refractor. UHC and OIII filters help to catch this one. If you can see NGC 7000, can you see the nearby Pelican Nebula, IC 5070?

Note that there is no picture available for this object, as it was invisible from the author's location without the aid of filtration.

NGC 7008

The short, dark lane running from the 13mag central star of this planetary nebula gives rise to its common name of the Fetus Nebula. It is best viewed using an OIII filter.

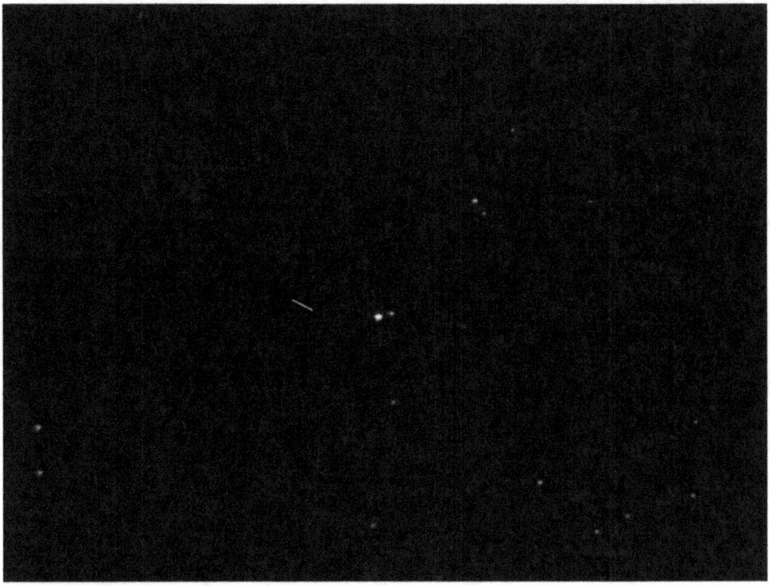

NGC 7008 11″ SCT

NGC 7008 72mm Refractor

NGC 7026

This small planetary nebula is sometimes referred to as the Cheeseburger Nebula, due to the slight, apparent darkening across the middle of its rectangular shape. High powers help to differentiate it from the surrounding field. Can you see the fainter nebulous extensions either side of the 'bun,' giving this planetary a more oval shape?

NGC 7026 11" SCT

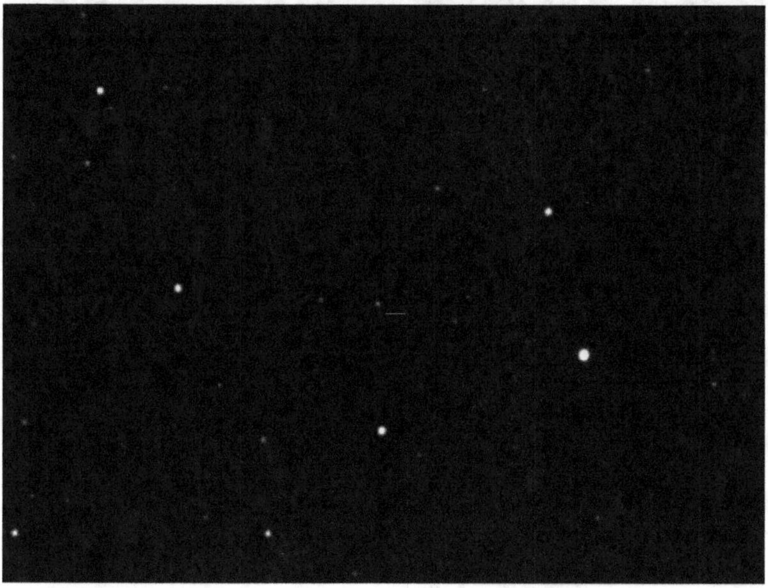

NGC 7026 72mm Refractor

NGC 7027

This is a small but bright planetary nebula with an 11mag star close by. At low powers it is very difficult to differentiate it from the surrounding field, and so an UHC or an OIII filter and/or high powers will help a great deal.

NGC 7027 11″ SCT

NGC 7063

This detached open cluster contains approximately 60 stars, although most will require very large apertures – hence the Trumpler Classification below.

Trumpler Classification: III 1 p

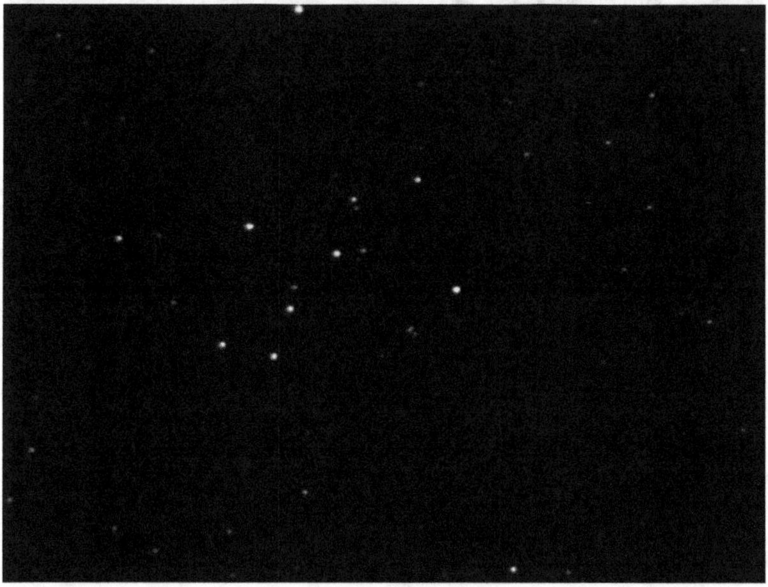

NGC 7063 11″ SCT

NGC 7063 72mm Refractor

Delphinus

Objects	Type	Position	Size	SB*
NGC 6934	GC	20h34′11.6″ +07 24 15	7.1′	
NGC 7006	GC	21h01′29.5″ +16 11 15	3.6′	

NGC 6934

This is a reasonably loose globular cluster, although the core is difficult to resolve. Can you resolve it?

NGC 6934 11″ SCT

NGC 6934 72mm Refractor

NGC 7006

This reasonably faint globular cluster has a bright center, which is very difficult to resolve.

NGC 7006 11″ SCT

NGC 7006 72mm Refractor

Draco

Objects	Type	Position	Size	SB*
M 102 or NGC 5866	Galaxy	15h06′29.7″ +55 45 46	6.4′×2.9′	12.9
NGC 5907	Galaxy	15h15′53.0″ +56 19 57	12.6′×1.4′	13.3
NGC 6503	Galaxy	17h49′29.2″ +70 08 28	7.1′×2.4′	13.2
NGC 6543	PN	17h58′33.5″ +66 37 59	20″	8.1

NGC 5866 or M 102?

This is a SA0sp galaxy that brightens gradually towards the center and has an oval shape. Using high magnification, can you see any mottling?

NGC 5866 11″ SCT

NGC 5866 72mm Refractor

NGC 5907

This is a SAc galaxy, presented edge-on. It brightens towards the central region. Can you make out the dust lane?

NGC 5907 11″ SCT

NGC 5907 72mm Refractor

NGC 6503

This is a SAcd galaxy with a 8.6mag star nearby. It has an oval-shaped halo, but the core is not well defined and appears 'smeared' across a good portion of the halo.

NGC 6503 11″ SCT

NGC 6503 72mm Refractor

NGC 6543

This planetary nebula is better known as the Cat's Eye Nebula, with a mag11 central star. Note the unusual apparently rectangular shape. An OIII filter really boosts the contrast. Can you make out any detail within the nebula?

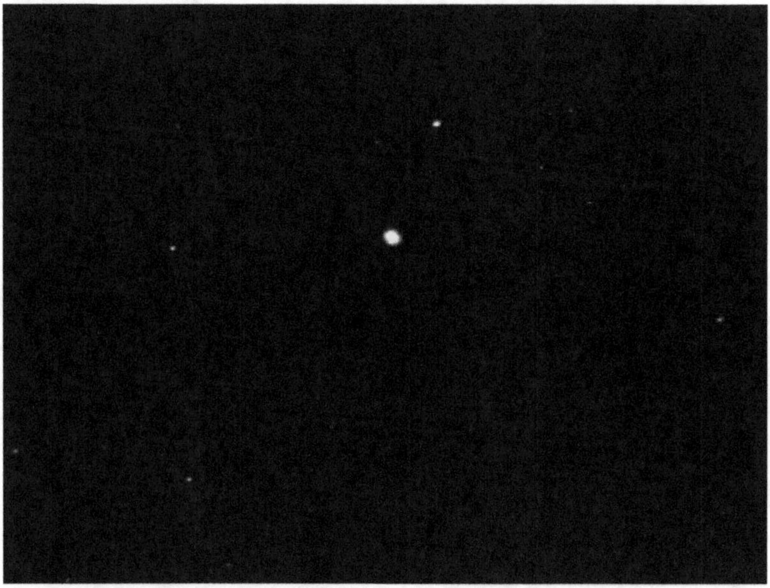

NGC 6543 11″ SCT

NGC 6543 72mm Refractor

Eridanus

Objects	Type	Position	Size	SB*
NGC 1232	Galaxy	03h09′45.1″ −20 34 47	7.4′×6.5′	13.9
NGC 1535	PN	04h14′15.8″ −12 44 21	51″	9.6

NGC 1232

This SABc galaxy has a very faint halo around a brighter core. Can you see any mottling (spiral arms) around it? A lot of detail in this galaxy is lost in the light pollution from the author's location, due to its low declination. It was invisible in the 72mm refractor.

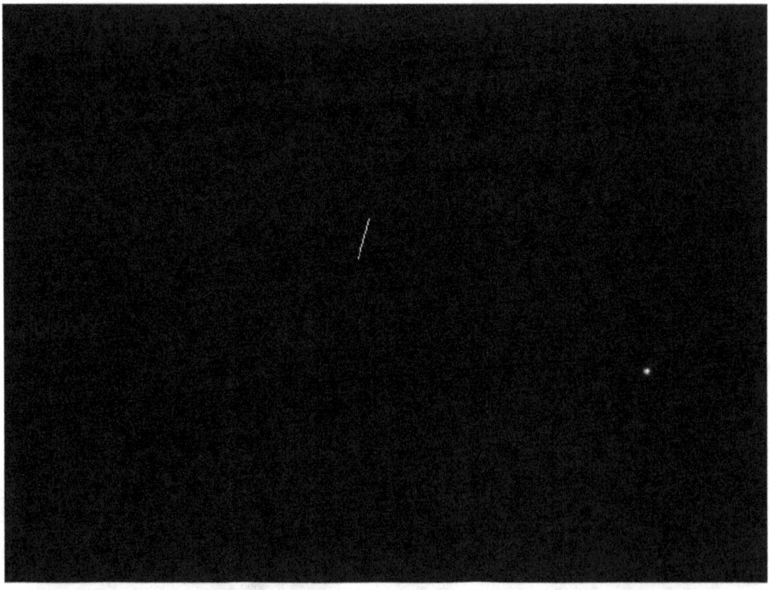

NGC 1232 11″ SCT

NGC 1535

This planetary nebula has a 12mag central star and is near a pair of 13/14mag stars. There is a hint of blue color. Can you see the detail (little 'tendrils' and lobes) within the disk (Hint: use high power)?

NGC 1535 11″ SCT

NGC 1535 72mm Refractor

Gemini

Objects	Type	Position	Size	SB*
IC 443	SN/EN	06h16′54.0″ +22 47 00	50′×40′	
M 35	OC	06h09′00.0″ +24 21 00	25′	
NGC 2129	OC	06h01′06.0″ +23 19 24	6′	
NGC 2158	OC	06h07′26.0″ +24 05 48	5′	
NGC 2266	OC	06h43′20.0″ +26 58 12	5′	
NGC 2304	OC	06h55′12.0″ +17 59 18	3′	
NGC 2355	OC	07h17′00.0″ +13 45 00	8′	
NGC 2371/2372	PN	07h25′34.8″ +29 29 22	62″	11.2
NGC 2392	PN	07h29′11.0″ +20 54 39	54″	9.1
NGC 2395	OC	07h27′13.0″ +13 37 00	15′	
NGC 2420	OC	07h38′24.0″ +21 34 24	6′	

IC 443

This is a large, diffuse supernova remnant. Part of it has a curious crescent-like shape with fine tendrils, and so the whole thing is sometimes referred to as the Jellyfish Nebula. This object is difficult to detect visually and would require the use of an UHC or an OIII filter. The brightest section (in the northeast) is just visible with averted vision in the bottom right-hand corner of the image below. If you can see this object, can you see the reflection nebula IC 444 nearby?

IC 443 11″ SCT

M 35

There are over 400 stars in this large open cluster. There is also a large range of brightness. Using low powers, can you detect NGC 2158 (see below)?

Trumpler Classification: III 3 r

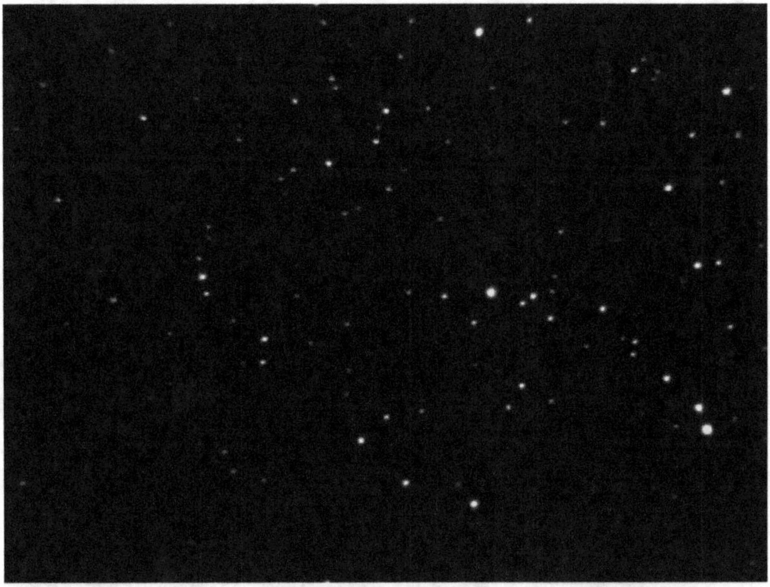

M 35 11″ SCT

M 35 72mm Refractor

NGC 2129

This is a moderately rich open cluster with stars ranging in brightness. It is well detached from the field and contains two stars that are obviously brighter than the rest.

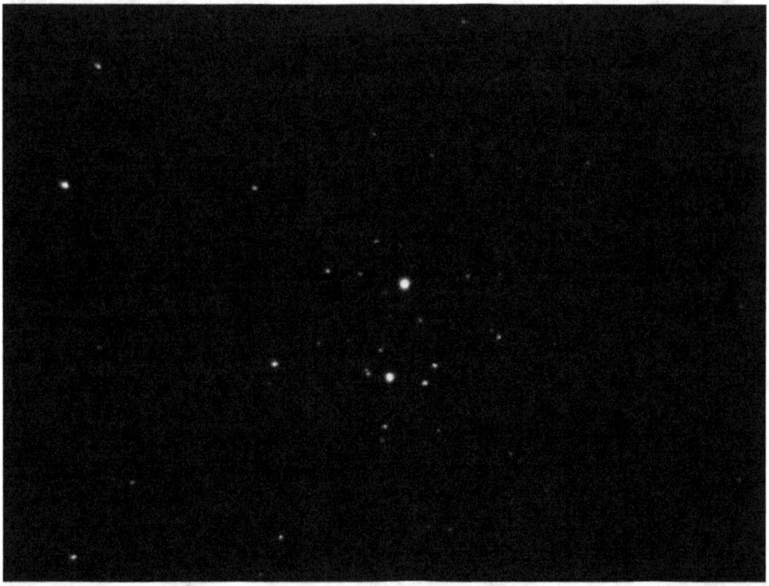

NGC 2129 11″ SCT

NGC 72mm Refractor

Trumpler Classification: I 3 m

NGC 2158

This cluster contains almost 1,000 stars, slightly concentrated towards the center. It is in the same low power field of view as M 35, although it is much dimmer, since it is much further away. M 35 can be seen in the top right of the 72mm refractor view below.

Trumpler Classification: II 3 r

NGC 2158 11″ SCT

NGC 2158 72mm Refractor

NGC 2266

There are approximately 50 stars in this cluster with a moderate range of brightness. This cluster is difficult in small apertures, as the majority of stars are fainter than mag11.

NGC 2266 11″ SCT

NGC 2266 72mm Refractor

Trumpler Classification: II 2 m

NGC 2304

This apparently small cluster contains about 30 stars of similar brightness.

Trumpler Classification: II 1 m/p

NGC 2304 11″ SCT

NGC 2304 72mm Refractor

NGC 2355

There are approximately 40 stars in this open cluster, with a moderate range of brightness.

NGC 2355 11″ SCT

NGC 2355 72mm Refractor

Trumpler Classification: II 2 m/p

NGC 2371 and 2372

This planetary nebula has two NGC numbers, one for each of the lobes. It is sometimes called the Peanut Nebula and responds well to an UHC or an OIII filter. The central star is at the very limit of large telescopes from suburban locations, shining at mag14.8. Can you see it?

NGCs 2371 and 2372 11″ SCT

NGCs 2371 and 2372 72mm Refractor

NGC 2392

This famous planetary is more commonly known as the Clown Face or the Eskimo Nebula. It has a bright central star, mag10.5. This nebula responds well to UHC or OIII filters and high magnification. Can you see the face-like shape within the 'furry hood'? (Hint: Use high power.)

NGC 2392 11″ SCT

NGC 2392 72mm Refractor

NGC 2395

There are about 60 stars in this cluster over a moderate range of brightness. It is not well detached from the surrounding field.

Trumpler Classification: IV 2 m

NGC 2395 11″ SCT

NGC 2395 72mm Refractor

NGC 2420

This cluster contains approximately 300 stars over a small range of brightness. It is well detached from the field and concentrated towards the center.

Trumpler Classification: I 1 r

NGC 2420 11″ SCT

NGC 2420 72mm Refractor

Hercules

Objects	Type	Position	Size	SB*
M 13	GC	16h41′41.5″ +36 27 37	20.0′	
M 92	GC	17h17′07.3″ +43 08 11	14.0′	
NGC 6210	PN	16h44′29.7″ +23 47 58	21″	

M 13

This globular cluster is often referred to as the Great Hercules Cluster. Can you see galaxy IC 4617, just to the north northeast (you'll need a large aperture)? Galaxy NGC 6207 is also visible in larger fields of view.

M 13 11″ SCT

M 13 72mm Refractor

M 92

This is a very attractive globular cluster and can be resolved to the core.

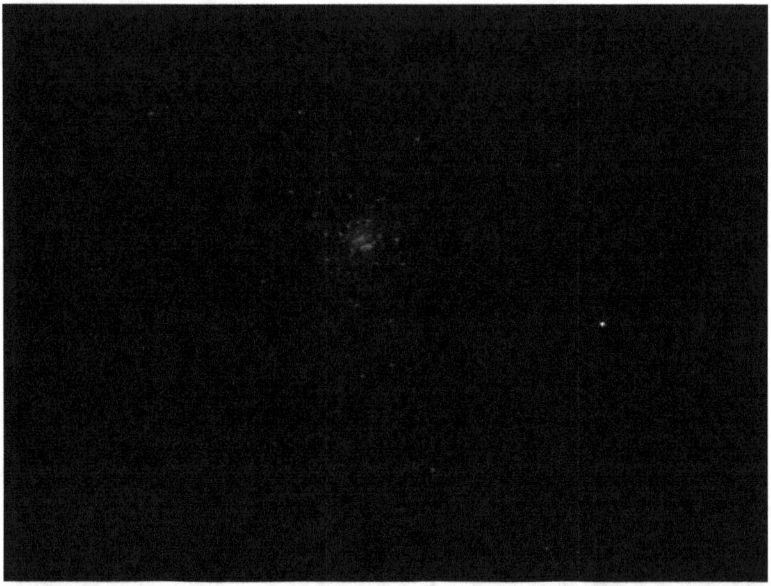

M 92 11″ SCT

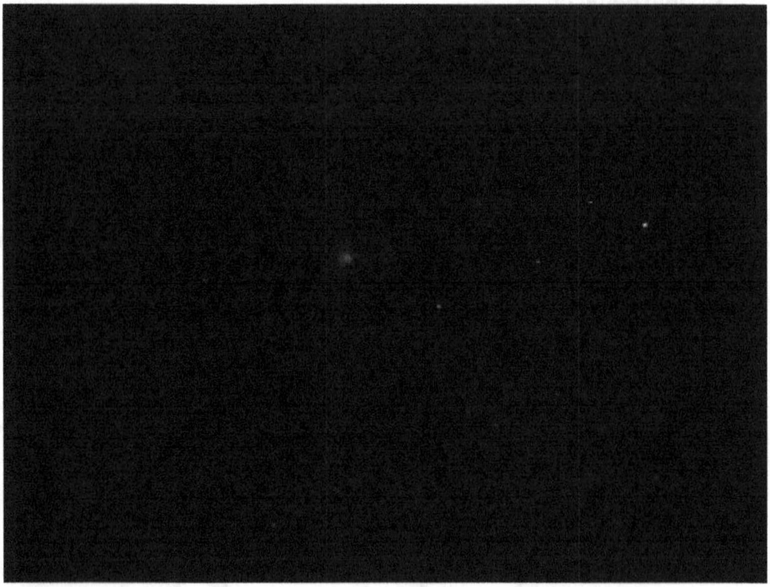

M 92 72mm Refractor

NGC 6210

This planetary nebula contains a bright 12mag central star and some tendrils that extend north and south. Can you see them (use high power)? Use an UHC or an OIII filter to help you.

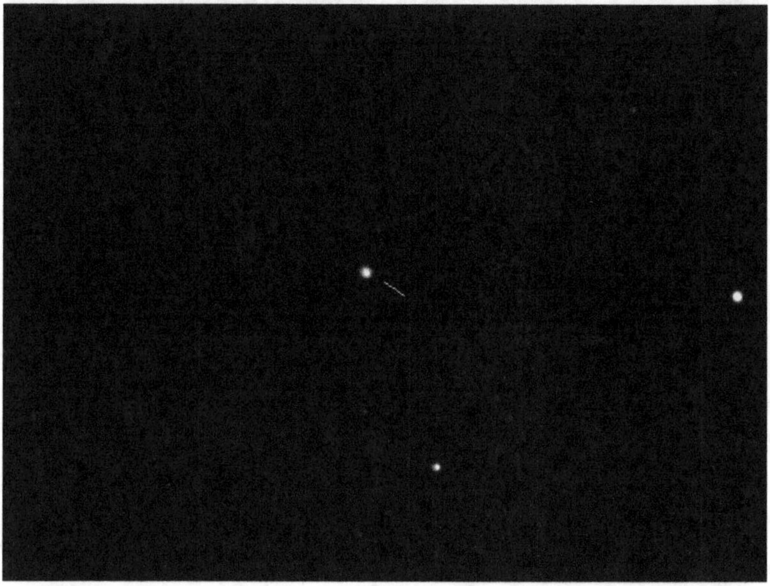

NGC 6210 11″ SCT

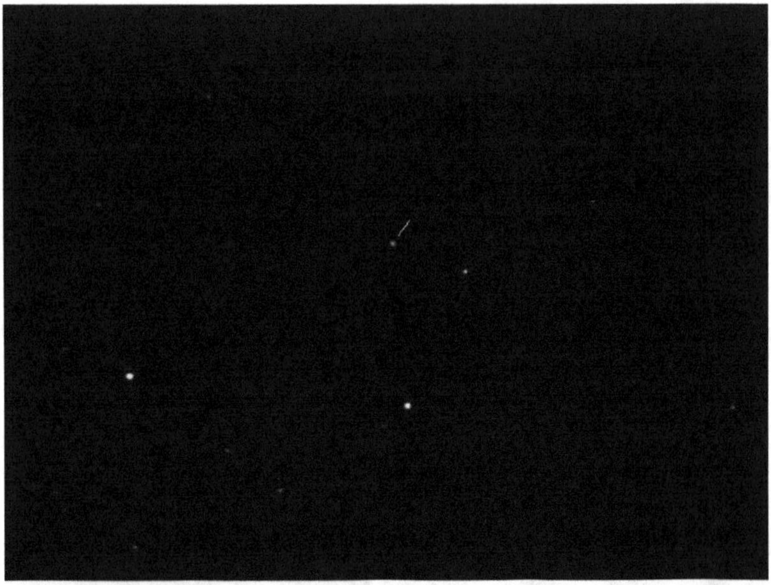

NGC 6210 72mm Refractor

Hydra

Objects	Type	Position	Size	SB*
M 48	OC	08h13′44.0″ −05 45 00	30′	
NGC 3242	PN	10h24′46.2″ −18 38 34	64″	

M 48

This apparently large open cluster contains around 80 stars, has a large range of brightness, and forms a triangular shape.

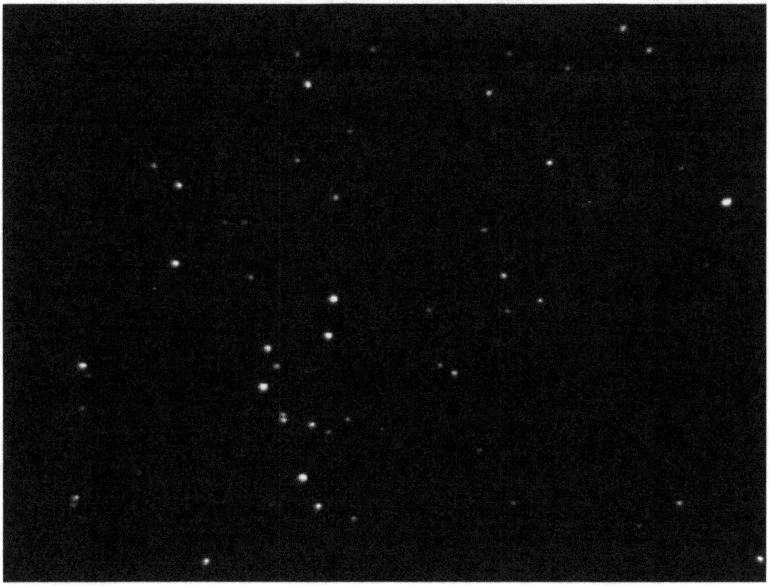

M 48 11″ SCT

M 48 72mm Refractor

Trumpler Classification: I 3 r

NGC 3242

This bright blue-green planetary is more commonly known as the Ghost of Jupiter, probably due to its angular size and brightness. The central star is a 'gettable' 13.3mag. Use high powers. Can you see the 'eye' looking back at you? Using an UHC or an OIII filter will help.

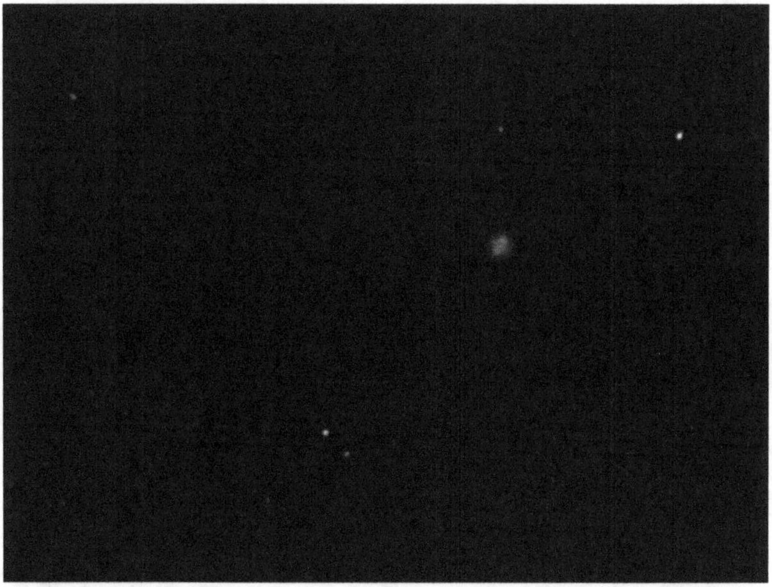

NGC 3242 11″ SCT

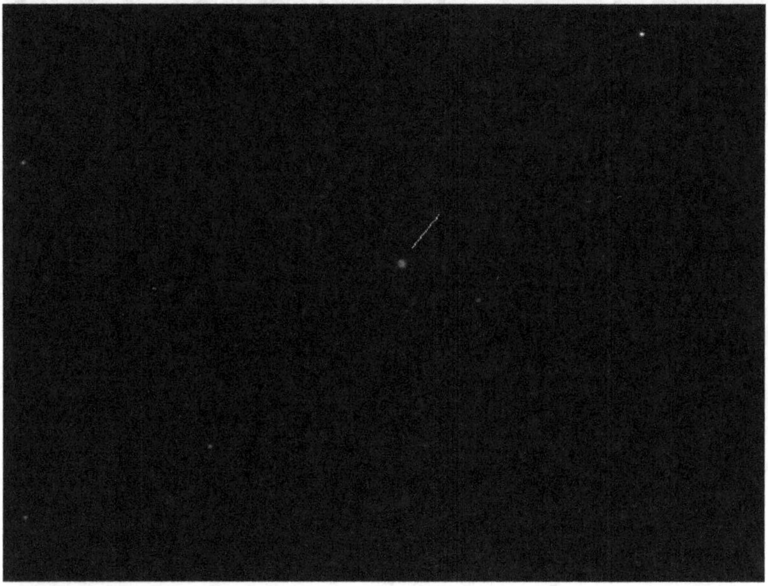

NGC 3242 72mm Refractor

Lacerta

Objects	Type	Position	Size	SB*
NGC 7209	OC	22h05′08.0″ +46 29 00	15′	
NGC 7243	OC	22h15′09.0″ +49 54 00	30′	

NGC 7209

This open cluster contains almost 100 stars and is similar in apparent size to NGC 7243.

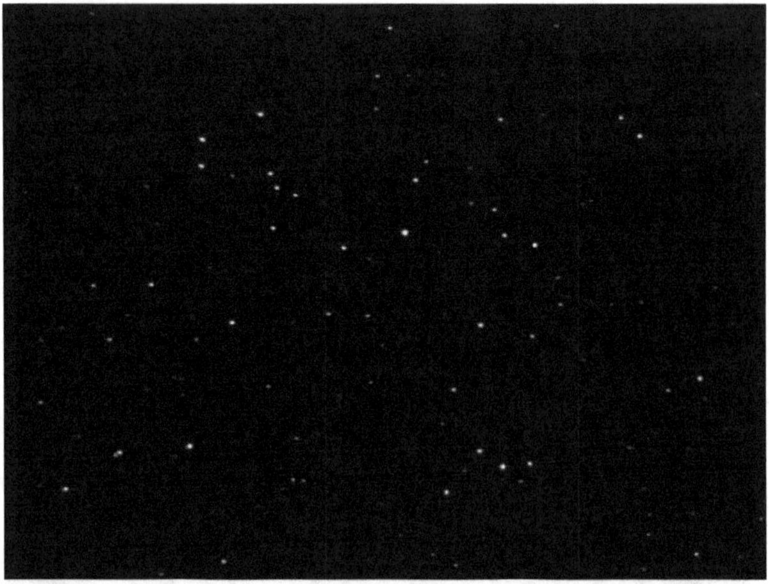

NGC 7209 11″ SCT

NGC 7209 72mm Refractor

Trumpler Classification: III 1 m/p

NGC 7243

This loose open cluster contains approximately 50 stars and is detached from the surrounding field.

Trumpler Classification: II 2 m

NGC 7243 11″ SCT

NGC 7243 72mm Refractor

Leo

Objects	Type	Position	Size	SB*
Abell 1367	Gx Cluster	11h44′30.0″ +19 50 00	100′	
M 65	Galaxy	11h18′55.7″ +13 05 37	9.8′×2.9′	12.8
M 66	Galaxy	11h20′15.4″ +12 58 59	9.1′×4.2′	12.7
M 95	Galaxy	10h43′57.4″ +11 42 15	7.4′×5.0′	13.5
M 96	Galaxy	10h46′45.9″ +11 49 12	7.6′×5.2′	13.1
M 105	Galaxy	10h47′49.8″ +12 34 55	5.4′×4.8′	12.8
NGC 2903	Galaxy	09h32′10.2″ +21 29 54	12.6′×6.0′	13.6
NGC 2964	Galaxy	09h42′54.3″ +31 50 48	2.9′×1.6′	12.8
NGC 3187	Galaxy	10h17′48.1″ +21 52 25	3.6′×1.6′	15.2
NGC 3190	Galaxy	10h18′05.9″ +21 49 59	4.4′×1.5′	13.0
NGC 3193	Galaxy	10h18′25.0″ +21 53 34	2.0′×2.0′	12.5
NGC 3384	Galaxy	10h48′16.9″ +12 37 44	5.5′×2.5′	12.6
NGC 3521	Galaxy	11h05′49.0″ −00 02 10	11.0′×5.5′	13.3
NGC 3628	Galaxy	11h20′18.2″ +13 35 06	14.8′×3.0′	13.4
NGC 3607	Galaxy	11h16′54.6″ +18 03 07	4.7′×3.9′	12.9
NGC 3753	Galaxy	11h37′53.9″ +21 58 49	1.7′×0.5′	13.3
NGC 3810	Galaxy	11h40′58.6″ +11 28 10	4.4′×3.0′	13.4

Abell 1367

This cluster of galaxies spans approximately 100′ and contains over 100 galaxies of varying brightness. (Look for the faint blobs in the image below.) How many can you find? Note that these objects were too dim to be visible in the small refractor.

Abell 1367 11″ SCT

M 65 and M 66

These two galaxies are classed SABa (M 65) and SABb (M 66) and form part of the famous Leo Triplet, with NGC 3628 very nearby. All three are visible in the same field of view using low powers, but NGC 3628 is very difficult in light polluted areas (see later).

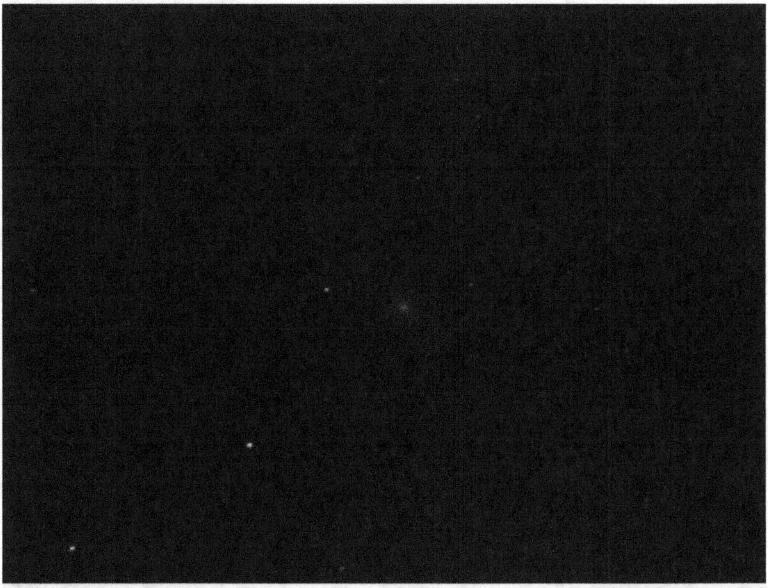

M 65 11″ SCT

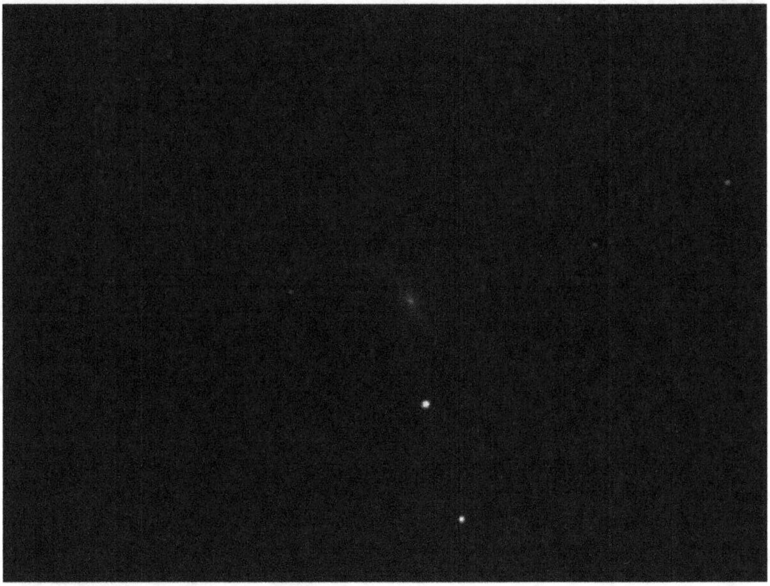

M 66 11″ SCT

M 65 and M 66 72mm Refractor

M 95

This galaxy is classed as a SBb and is located close to M 96 – in the same low power field of view. Is that a hint of a bar visible?

M 95 11" SCT

M 95 72mm Refractor

M 96

This galaxy is classed as a SABab and can be found close to M 95 (see above). Can you detect any structure in the halo?

M 96 11″ SCT

M 96 72mm Refractor

M 105 and NGC 3384

M 105 is an E1 galaxy that forms a nice pair with NGC 3384 – a SB0 galaxy visible in the center in both of the images below. Galaxy NGC 3389 is just out of the field of view (top-center) in the 11″ SCT image below.

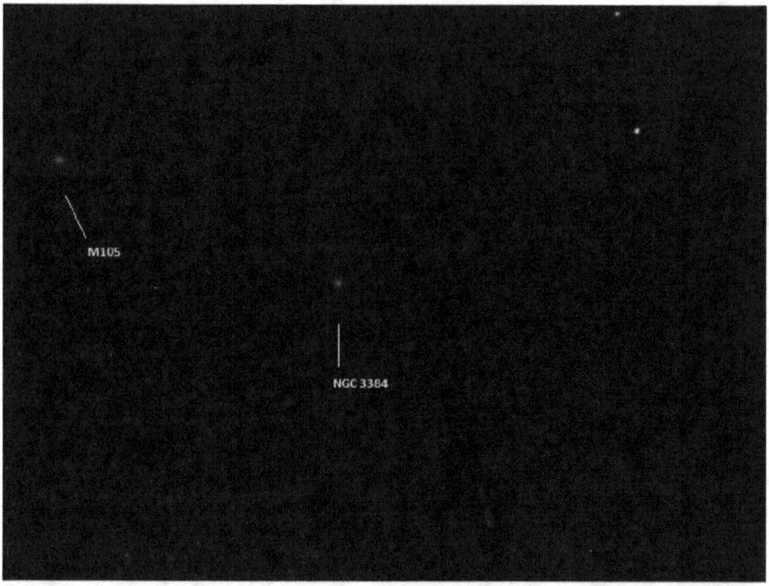

M 105 and NGC 3384 11″ SCT

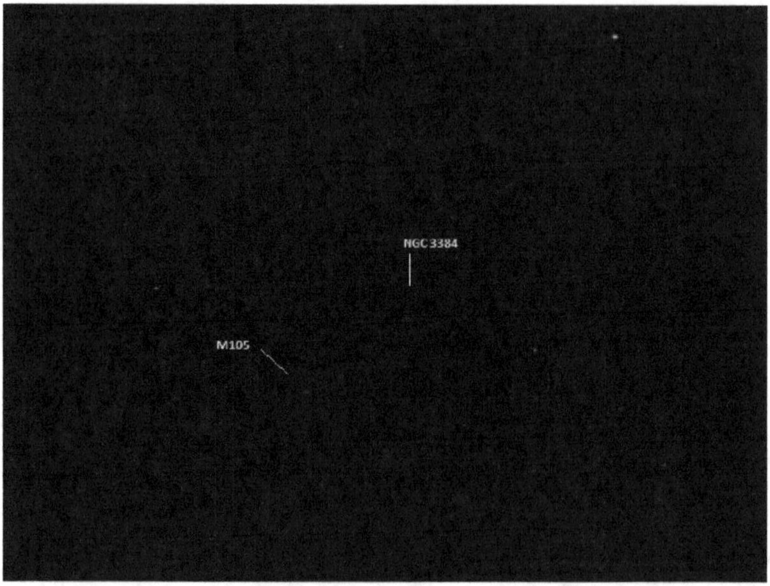

M 105 and NGC 3384 72mm Refractor

NGC 2903

This is a SABbc galaxy presented to us face-on and contains some detail in the halo. Can you see the knots and mottling?

NGC 2903 11″ SCT

NGC 2903 72mm Refractor

NGC 2964

This is a SABbc galaxy with a stellar nucleus and faint halo. Is there a hint of structure there?

NGC 2964 11" SCT

NGC 2964 72mm Refractor

NGC 3187, 3190, and 3193

All of these galaxies can be found in same field of view and form part of the 'Hickson 44' galaxy cluster. They are classed SBc pec, SAa pec, and E2, respectively. NGC 3187 is extremely faint and *just* perceptible with averted vision. Can you make out the dust lane along the length of NGC 3190?

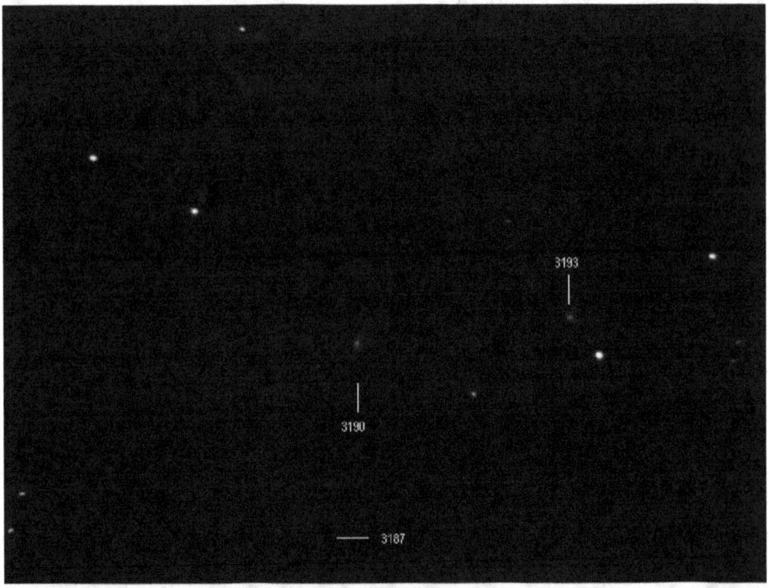

NGC 3187, 3190, and 3193 11″ SCT

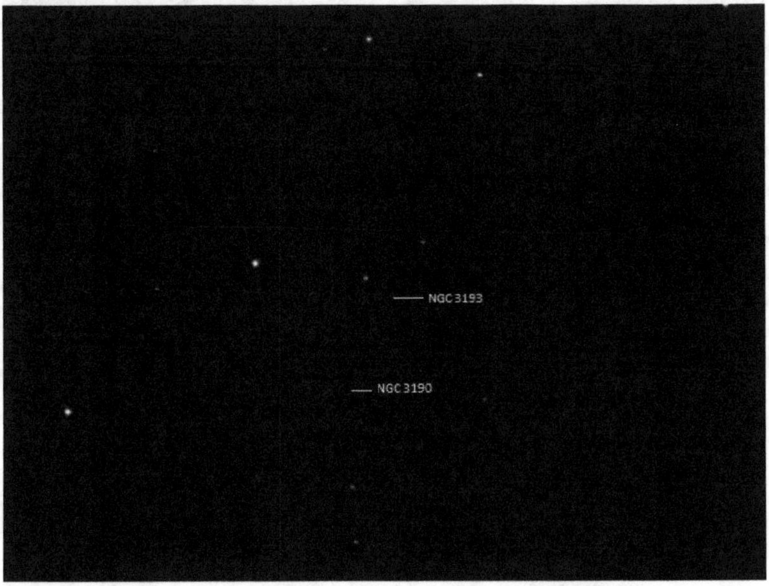

NGC 3190 and 3193 72mm Refractor

NGC 3521

This galaxy is classed as a SABbc and has a stellar nucleus. At higher powers, some mottling is visible near the core and in some of the bright areas around it. Is there a dust lane on the western side?

NGC 3521 11″ SCT

NGC 3521 72mm Refractor

NGC 3607

This is a SA0 galaxy with NGC 3605 nearby. Can you make out any structure?

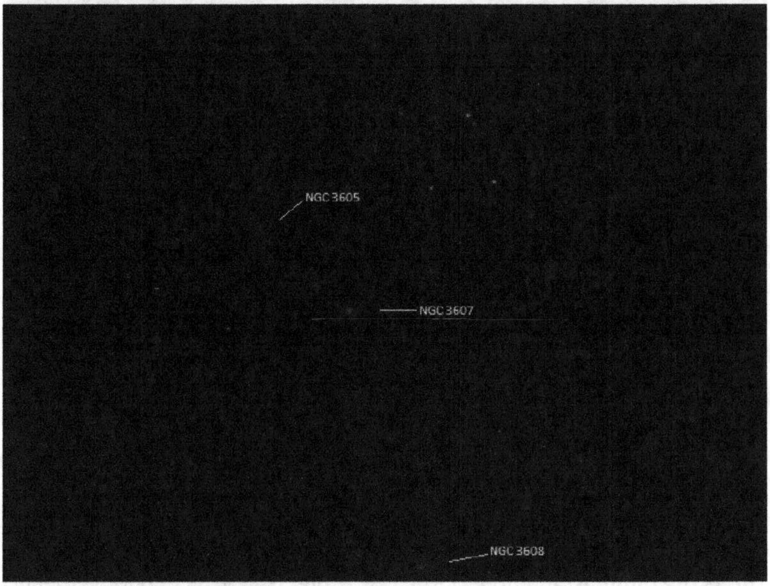

NGCs 3607, 3605, and 3608 11″ SCT

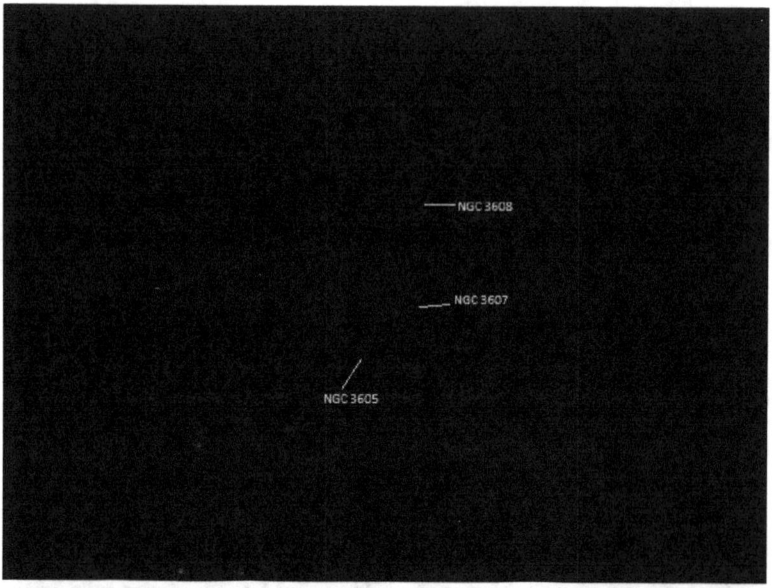

NGCs 3607, 3605, and 3608 72mm Refractor

NGC 3628

This is the dimmest member of the Leo Triplet (see M 65 and M 66 earlier) and is difficult to see from a suburban location. In the image below, you can just make it out between the two white lines using averted vision. Note the thick dust lane bisecting both dim parts. This object was invisible in the small refractor.

NGC 3628 11" SCT

NGC 3651

This class E galaxy is otherwise known as Hickson 51A and is one of a group of seven galaxies: Hickson 51 A, B, C, D, E, F, and G. Hickson 51C is also designated NGC 3653, and Hickson 51B is also IC 2759. How many can you make out?

Hickson 51 (including NGC 3651) 11″ SCT

NGC 3651 72mm Refractor

NGC 3753

This Sab galaxy is also known as Hickson 57A and forms part of the famous galaxy cluster, Copeland's Septet. How many of its friends can you make out? Note that this object was invisible in the small refractor.

Copeland's Septet 11″ SCT

NGC 3810

This is a Sac galaxy with some knots visible. How much of the halo can you make out?

NGC 3810 11″ SCT

NGC 3810 72mm Refractor

Leo Minor

Objects	Type	Position	Size	SB
NGC 3003	Galaxy	09h48′35.3″ +33 25 13	5.8′×1.3′	13.9
NGC 3344	Galaxy	10h43′31.3″ +24 55 23	7.1′×6.5′	13.9
NGC 3432	Galaxy	10h52′32.1″ +36 37 24	6.8′×1.5′	13.6

NGC 3003

This is a Sbc galaxy presented to us edge-on. It brightens slightly towards the center. How much of the halo can you see?

GC 3003 11″ SCT

NGC 3003 72mm Refractor

NGC 3344

This is a SABbc galaxy with a bright central region and a stellar nucleus. Note the two mag10 stars on the eastern edge. Can you make out any structure in the halo (barely visible 'arms' in the 11″ SCT view below)?

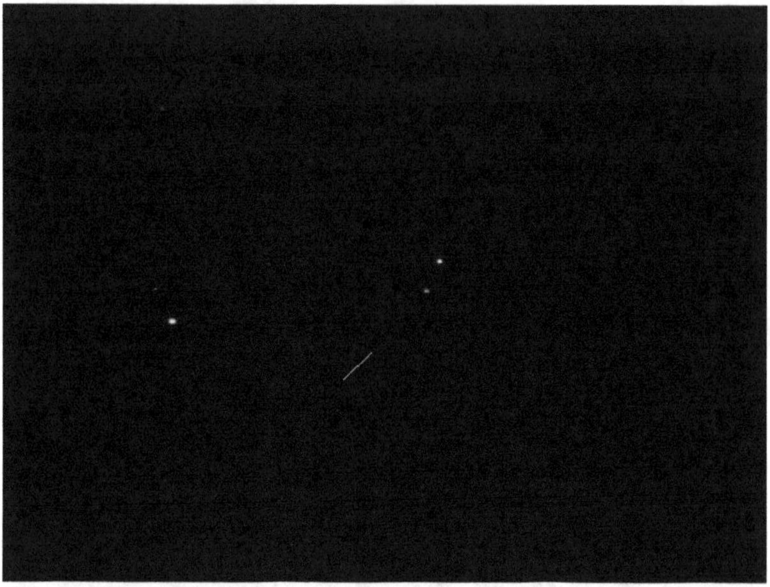

NGC 3344 11″ SCT

NGC 3344 72mm Refractor

NGC 3432

This galaxy is classed as a SBm sp with a mag9 star close by as well as UGC 5983. It is presented edge-on to us. How much of the halo can you see? Does it extend to the optical binary close by? Note the variation in brightness of the halo from one end to the other.

NGC 3432 11″ SCT

NGC 3432 72mm Refractor

Lepus

Objects	Type	Position	Size	SB
NGC 1784	Galaxy	05h05'27.2" −11 52 17	4.0'×2.5'	14.0

NGC 1784

This very dim SBc galaxy appears to show a few superimposed stars. Are you able to make out any more of the halo? As you can see, it is just visible in the 11" SCT picture below and barely perceptible in the 72mm refractor view (use averted vision).

NGC 1784 11" SCT

NGC 1784 72mm Refractor

Libra

Objects	Type	Position	Size	SB
NGC 5792	Galaxy	14h58′23.0″ −01 05 28	6.9′×1.7′	13.8

NGC 5792

This faint SBb galaxy has a mag9.6 star (left of center in the picture below) just 1′ from its nucleus.

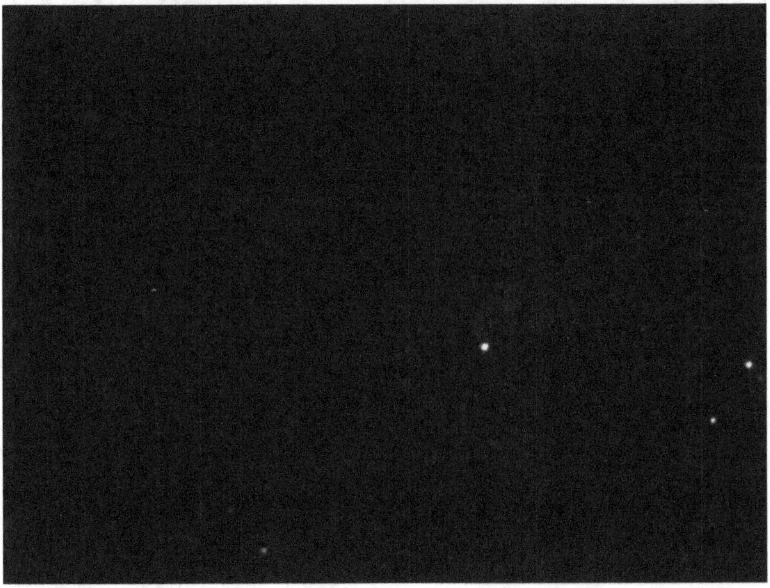

NGC 5792 11″ SCT

NGC 5792 72mm Refractor

Lynx

Objects	Type	Position	Size	SB
NGC 2683	Galaxy	08h52′40.2″ +33 25 01	9.3′×2.1′	12.8

NGC 2683

This galaxy, classed as a SAb, is presented to us edge-on with a brighter central region. Can you make out the mottling next to the core?

NGC 2683 11″ SCT

NGC 2683 72mm Refractor

Lyra

Objects	Type	Position	Size	SB
M 56	GC	19h16′35.5″ +30 11 05	8.8′	
M 57	PN	18h53′35.2″ +33 01 44	76″	

M 56

This loose globular cluster begins to be resolved in medium to high apertures, with about twenty stars visible within an irregularly shaped, 'nebulous' core. Can you see any dark lanes or other fine details?

M 56 11″ SCT

M 56 72mm Refractor

M 57

This famous planetary nebula is better known as the Ring Nebula. The 'hole' in the annular ring is best viewed through small aperture 'scopes, as larger apertures tend to fill the 'space' with faint nebulous material. An UHC filter will bring out some of the outer detail. This is a beautiful object on its own, but can you see galaxy IC 1296 only 4' away (this is quite a challenge)? Can you see the central star of M 57?

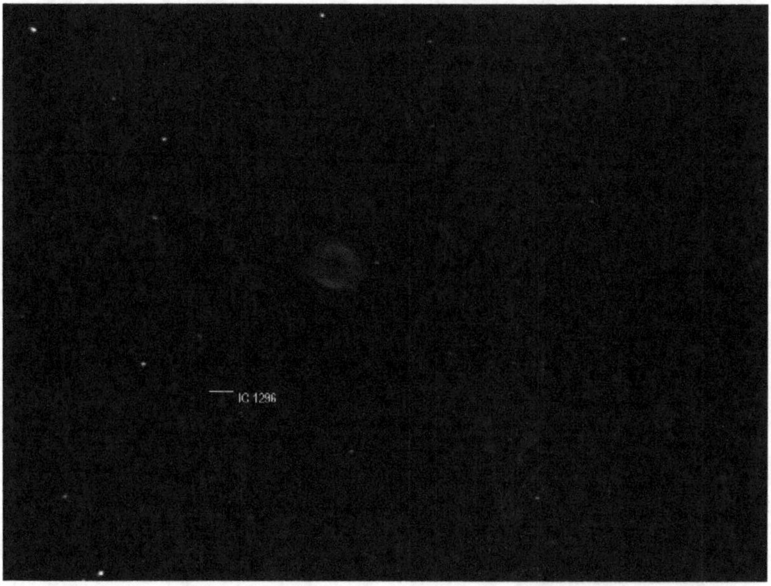

M 57 11" SCT

M 57 72mm Refractor

Monoceros

Objects	Type	Position	Size	SB*
M 50	OC	07h02′47.8″ −08 23 00	15′	
NGC 2232	OC	06h28′02.0″ −04 51 00	45′	
NGC 2237–2239	OC+EN	06h32′18.0″ +05 03 00	80′×60′	
NGC 2244	OC+EN	06h32′19.0″ +04 51 00	~30′	
NGC 2251	OC	06h34′39.0″ +08 22 00	10′	
NGC 2261	EN/RN	06h39′12.0″ +08 44 00	3.4′×1.5′	
NGC 2264	OC+EN	06h40′59.0″ +09 54 00	40′+35′×15′	
NGC 2301	OC	06h51′46.0″ +00 28 00	15′	
NGC 2343	OC	07h08′07.0″ −10 37 00	6′	

M 50

This open cluster contains approximately 80 stars with a large range in brightness and some concentration towards the center.

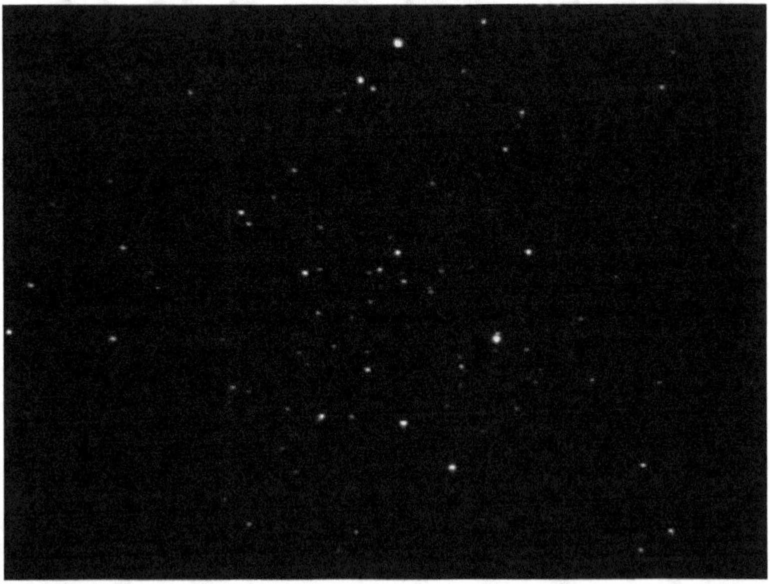

M 50 11″ SCT

M 50 72mm Refractor

Trumpler Classification: II 3 r/m

NGC 2232

There are very few stars over a wide area in this open cluster. Note that the star 10-Mon is near the center of the cluster.

Trumpler Classification: III 2 p

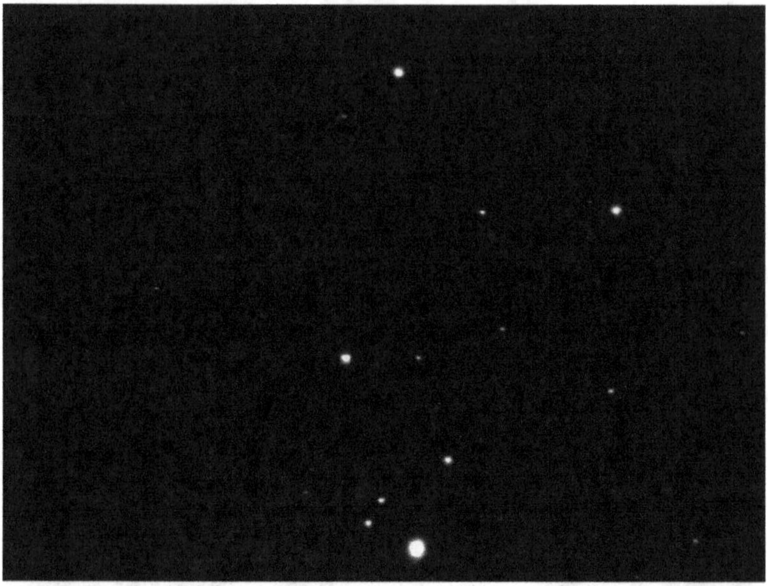

NGC 2232 11″ SCT

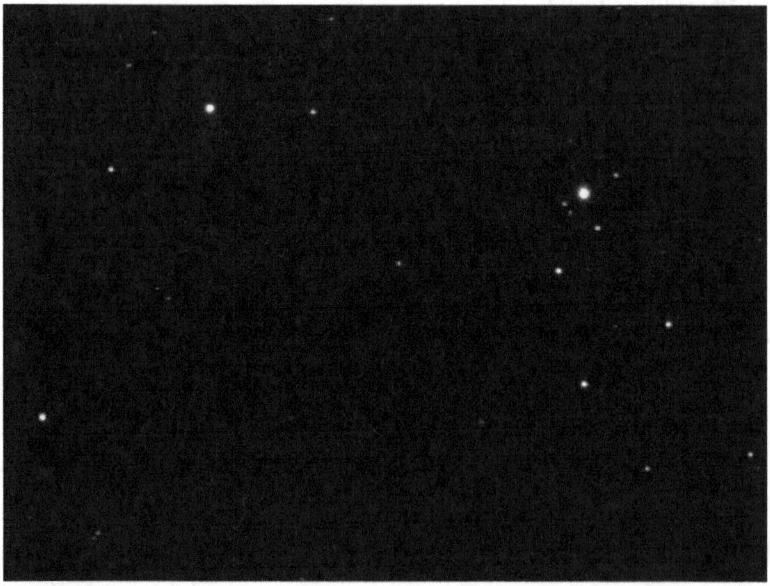

NGC 2232 72mm Refractor

NGC 2237–2239

This large area of nebulosity is better known as the Rosette Nebula, and it surrounds the open cluster NGC 2244. This object is best viewed using low powers and wide fields of view. An OIII or a an UHC filter will help bring out the nebula and extra detail. In a small refractor the nebula appears horseshoe-shaped. How much of the nebula can you see with a small refractor? Below is part of the field using the 72mm refractor. Note that the nebula is invisible.

NGC 2237–2239 72mm Refractor

NGC 2244

This is the cluster within the Rosette Nebula and contains approximately 100 stars with a large range of brightness. The brightest star is not a member of the cluster.

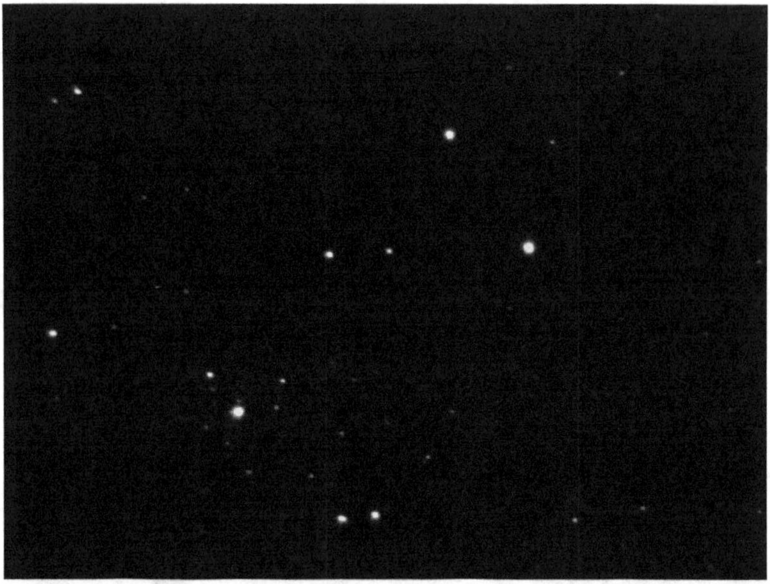

NGC 2244 11″ SCT

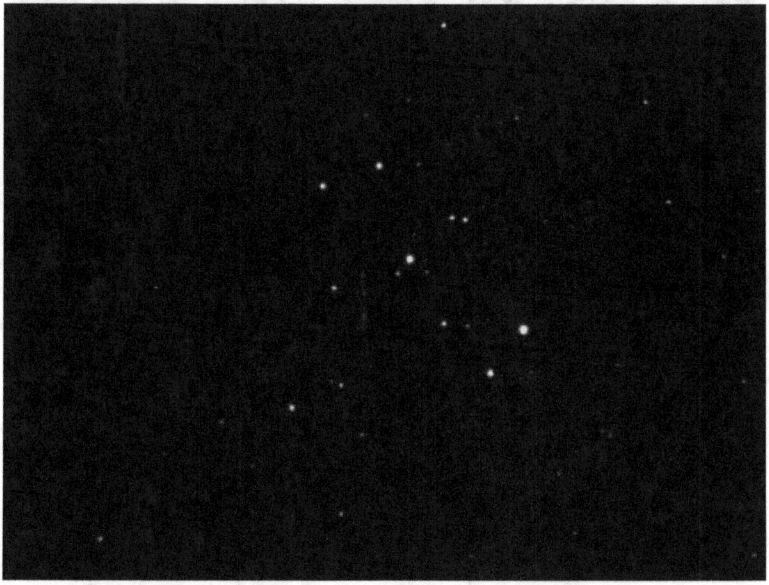

NGC 2244 72mm Refractor

Trumpler Classification: II 3 r n

NGC 2251

This open cluster contains almost 100 stars over a moderate range of brightness. There is no central concentration, but the stars appear almost 'grouped' in three sections.

Trumpler Classification: II/III 2 m

NGC 2251 11″ SCT

NGC 2251 72mm Refractor

NGC 2261

This bright nebula contains both emission and reflection components and is better known as Hubble's Variable Nebula. Note the distinctive fan shape with a bright star at the tip! Try high power on this object for extra detail. The seeing greatly affects the extent of the visible nebulosity as, indeed, does the nebula's variability in brightness!

NGC 2261 11″ SCT

NGC 2261 72mm Refractor

NGC 2264

This is a region of bright and dark nebulosity together with the open cluster named the Christmas Tree. It also contains the dark nebula called the Cone Nebula. Can you see the 'cone' pushing into the brighter part of the nebulosity near S Mon (an UHC filter will help)?

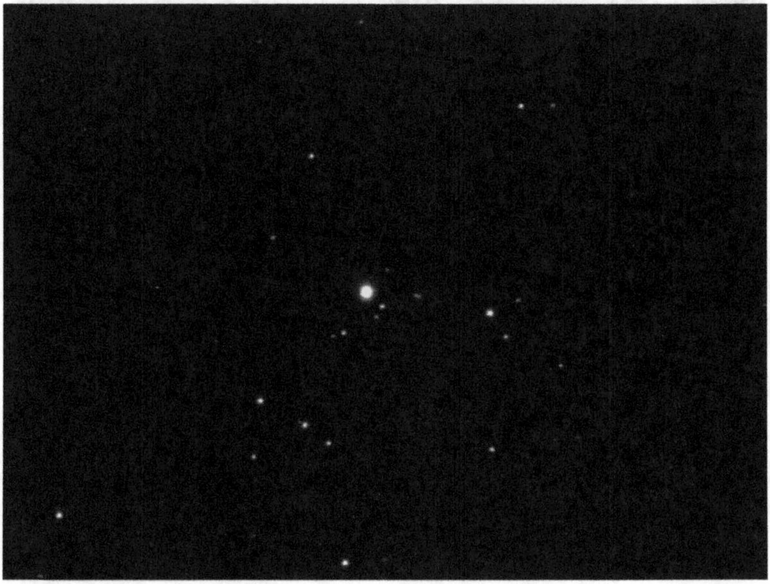

NGC 2264 72mm Refractor

Trumpler Classification: III 3 m n

NGC 2301

This open cluster contains 80 stars over a large range of brightness with a strong central concentration to an 8′ core.

Trumpler Classification: I 3 m/r

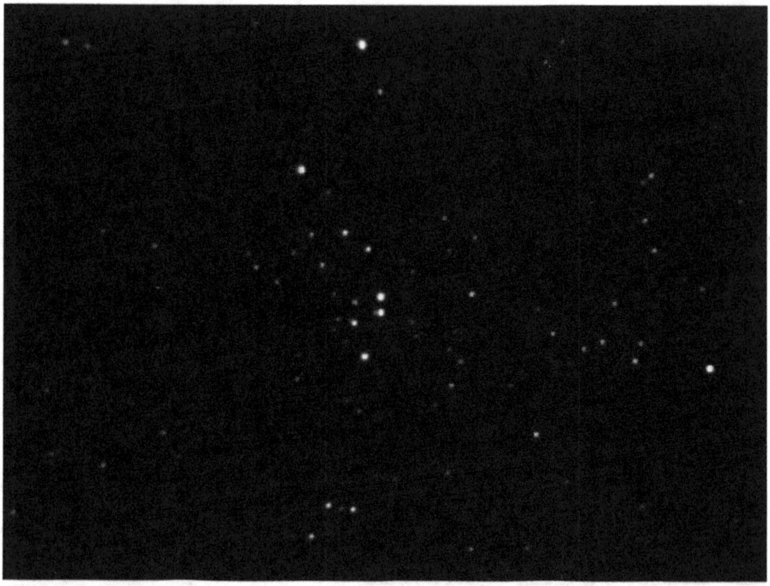

NGC 2301 11″ SCT

NGC 2301 72mm Refractor

NGC 2343

There are almost 50 stars over a moderate range of brightness in this cluster. There is a slight concentration towards the center. This cluster is embedded in nebulosity. Can you see it?

Trumpler Classification: II 2 p n

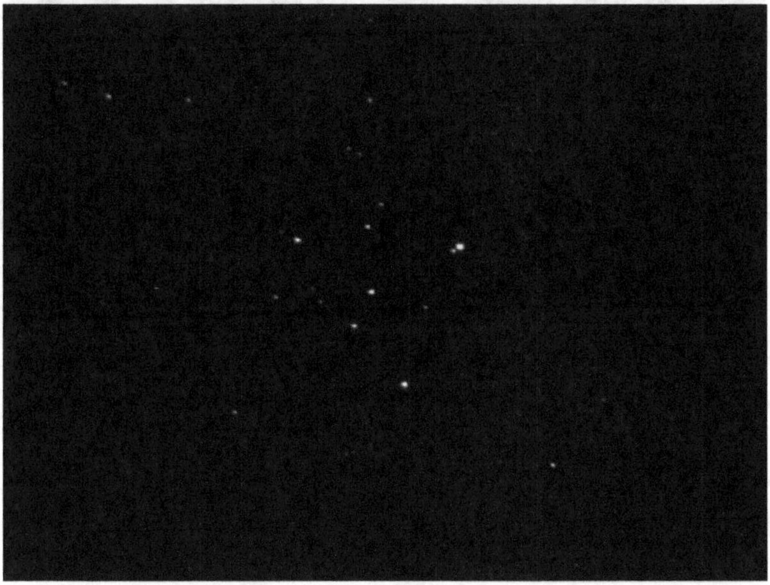

NGC 2343 11″ SCT

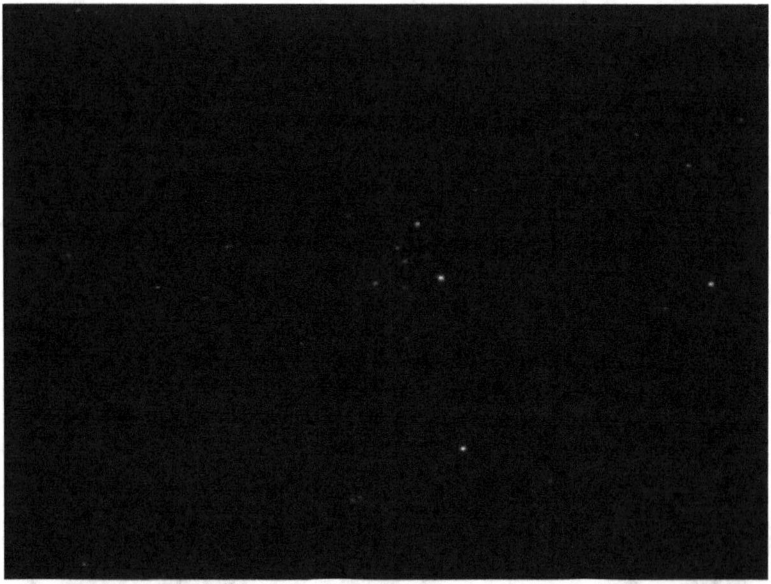

NGC 2343 72mm Refractor

Ophiuchus

Objects	Type	Position	Size	SB*
IC 4665	OC	17h46′12.0″ +05 43 00	70′	
M 9	GC	17h19′11.8″ −18 30 59	12.0′	
M 10	GC	16h57′08.9″ −04 05 58	20.0′	
M 12	GC	16h47′14.5″ −01 56 52	16.0′	
M 14	GC	17h37′36.1″ −03 14 45	11.0′	
M 107	GC	16h32′31.9″ −13 03 13	13.0′	
NGC 6572	PN	18h12′06.4″ +06 51 11	15″	
NGC 6633	OC	18h27′15.0″ +06 30 00	20′	

IC 4665

This open cluster contains over 50 stars with a moderate range of brightness. There is no concentration towards the center, and it requires large fields of view (binoculars?) to fully appreciate it!

Trumpler Classification: III 2 m

IC 4665 11″ SCT

IC 4665 72mm Refractor

M 9

This is a relatively faint globular cluster with a slightly irregular (non-circular) look. Note the chains of mag11 to mag13 stars.

M 9 11″ SCT

M 10

This is a fairly bright and large globular cluster. It appears slightly dimmer than its close friend M 12. Can you resolve the core?

M 10 11″ SCT

M 10 72mm Refractor

M 12

This globular cluster is of a similar size and brightness to nearby M 10. At a glance it appears similar in shape to M 13 in Hercules.

M 12 11″ SCT

M 12 72mm Refractor

M 14

This globular cluster is quite faint and similar in brightness to M 9. However, this cluster can be resolved at higher powers.

M 14 11″ SCT

M 14 72mm Refractor

M 107

This is a very loose and faint globular cluster that could, by its appearance, be mistaken for an open cluster.

M 107 11″ SCT

NGC 6572

A 'powerful' central star gives this planetary nebula an intense blue-green color in large apertures. Note how this object appears stellar in the images below.

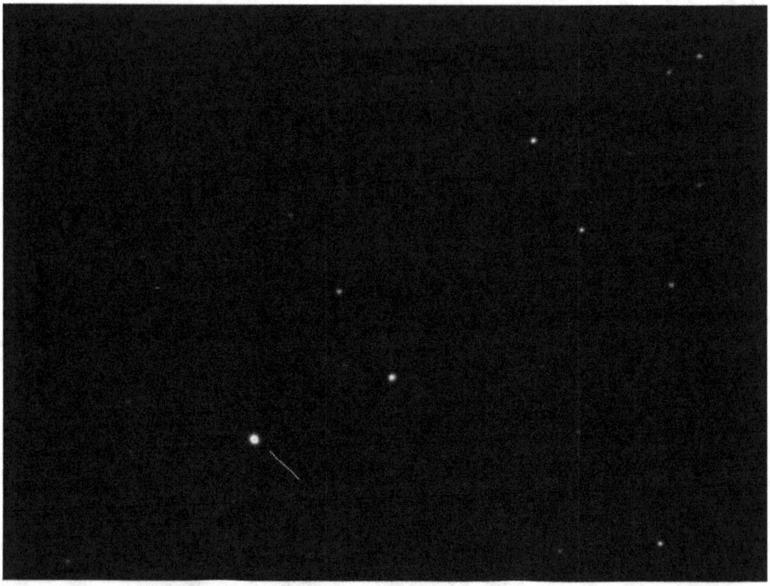

NGC 6572 11″ SCT

NGC 6572 72mm Refractor

NGC 6633

This large open cluster contains over 150 stars with a moderate range of brightness (many are undetectable in amateur equipment). As you can see, there is no condensation towards the center, and it is not well detached from the field. This object is best viewed at low powers.

Trumpler Classification: III 2 m

NGC 6633 11″ SCT

NGC 6633 72mm Refractor

Orion

Objects	Type	Position	Size	SB*
M 42/43	EN	05h35′24.0″ −05 27 00	65′×60′	
M 78	RN	05h46′42.0″ +00 03 00	8′×6′	
NGC 1662	OC	04h48′29.0″ +10 56 00	12′	
NGC 1788	RN	05h06′54.0″ −03 21 00	5×3′	
NGC 1973/5	EN/RN	05h35′06.0″ −04 44 00	5′×5′	
NGC 1981	OC	05h35′12.0″ −04 26 00	28′	
NGC 2022	PN	05h42′06.2″ +09 05 13	39″	11.6
NGC 2024	EN	05h41′54.0″ −01 51 00	30′×30′	
NGC 2194	OC	06h13′46.0″ +12 48 00	9′	

M 42/43

These two objects make up the famous Orion Nebula. M 43 is actually a part of M 42 but separated from it by a dark lane of dust. This is a large emission nebula with a great deal of detail. Alternating between a variety of filters provides extra detail around the central 'trapezium' region.

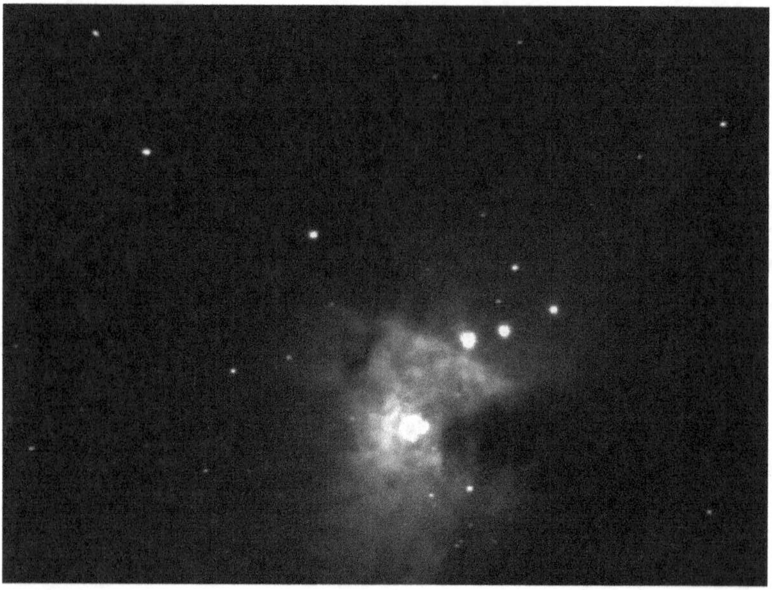

M 42 11″ SCT

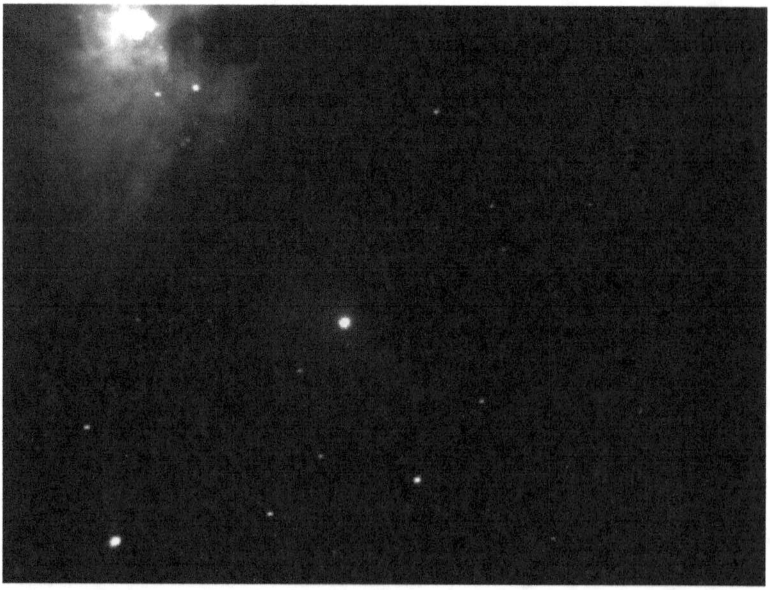

M 43 11″ SCT

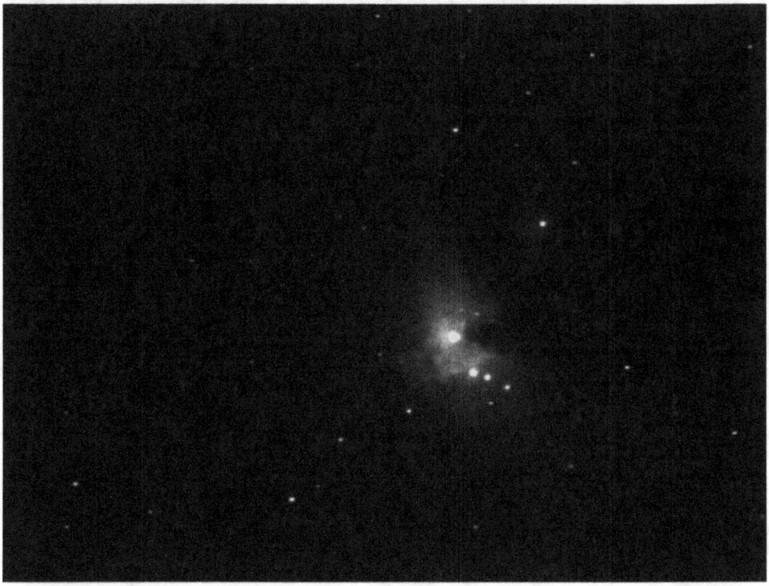

M 42 and M 43 72mm Refractor

M 78

This reflection nebula is the brightest of four nebulae – this one and NGCs 2064, 2067, and 2071. It contains two mag10 stars and appears almost comet-shaped due to a dark lane on one side. Can you make out the rest of the nebulosity (i.e. the other NGCs)? You'll need a large aperture.

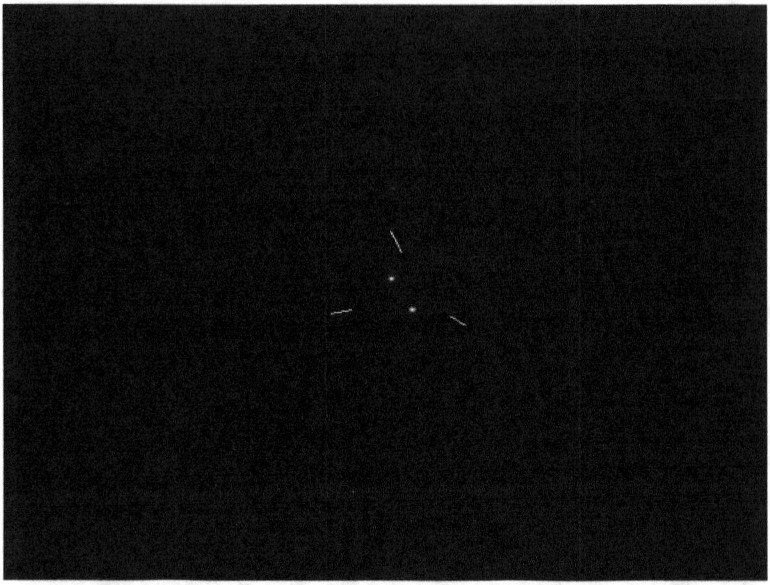

M 78 11″ SCT

M 78 72mm Refractor

NGC 1662

This large open cluster contains over 50 stars, although a significant number of these are dimmer than mag13.

Trumpler Classification: II 3 m

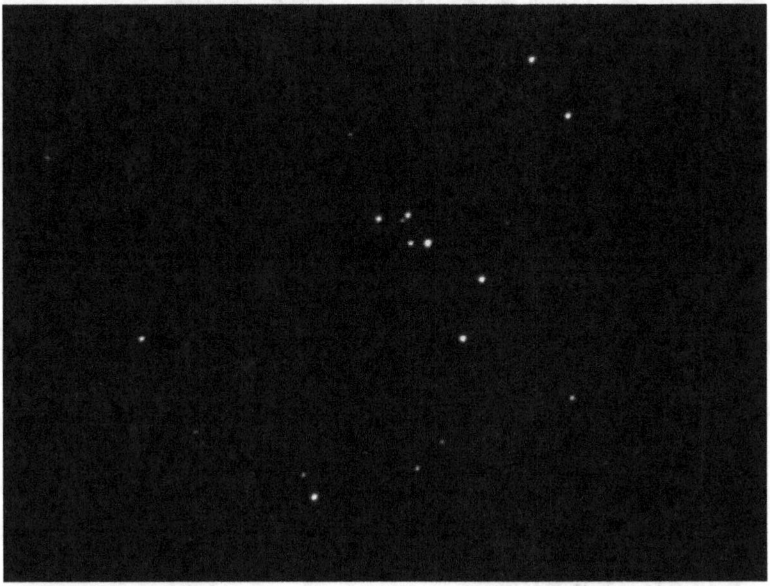

NGC 1662 11″ SCT

NGC 1662 72mm Refractor

NGC 1788

This is a very faint reflection nebula and has the dark nebula LDN 1616 cutting into it. The brightest star involved is mag10 and lies in the northwest section. How much nebulosity can you make out? This object was invisible in the author's small refractor.

NGC 1788 11″ SCT

NGCs 1973, 1975, and 1977

These are really one reflection nebula associated with KX Orionis. They have three separate NGC numbers as the sections are divided by dust lanes making a shape reminiscent of a 'Running Man' – hence the popular name. As these types of nebulae go, it is reasonably bright, but an UHC filter is recommended in order to glean more detail and nebulosity.

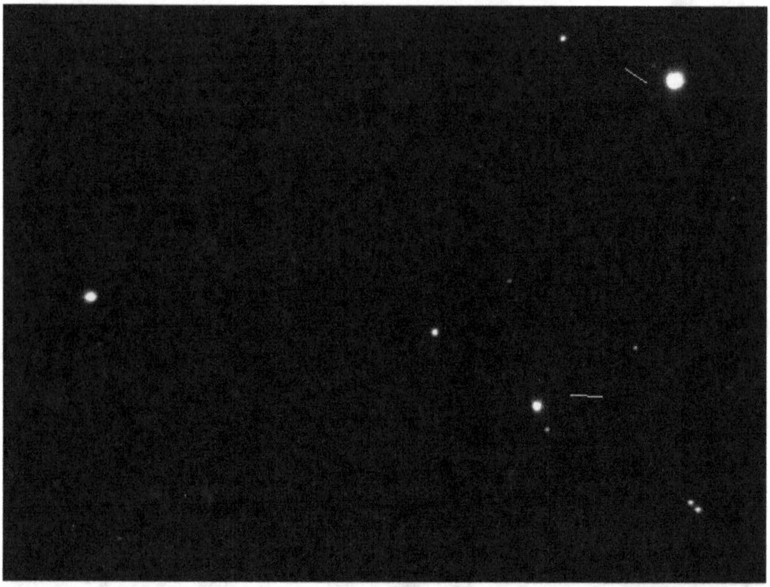

NGCs 1973 and 1975 11″ SCT

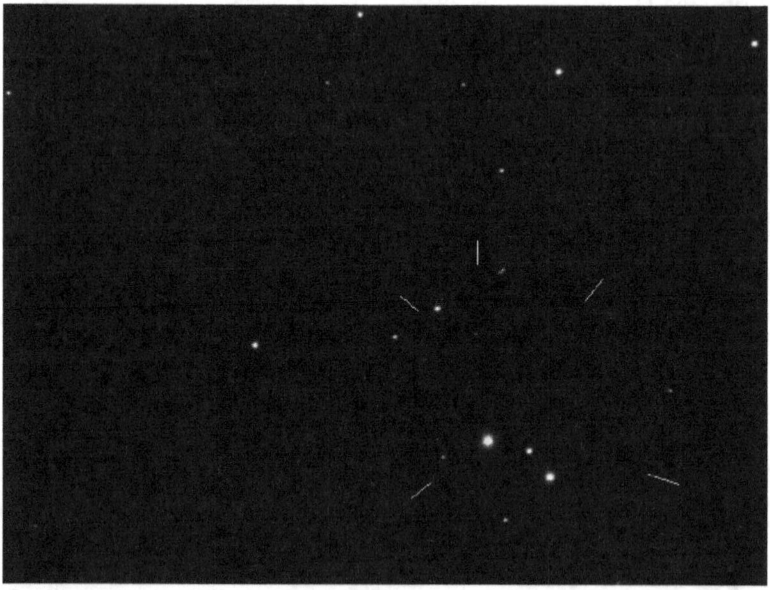

NGCs 1973 and 1975 72mm Refractor

NGC 1981

This open cluster contains about 20 stars over a large range of brightness. Can you detect the nebulosity in the south from NGCs 1973–1977?

Trumpler Classification: III 3 p n

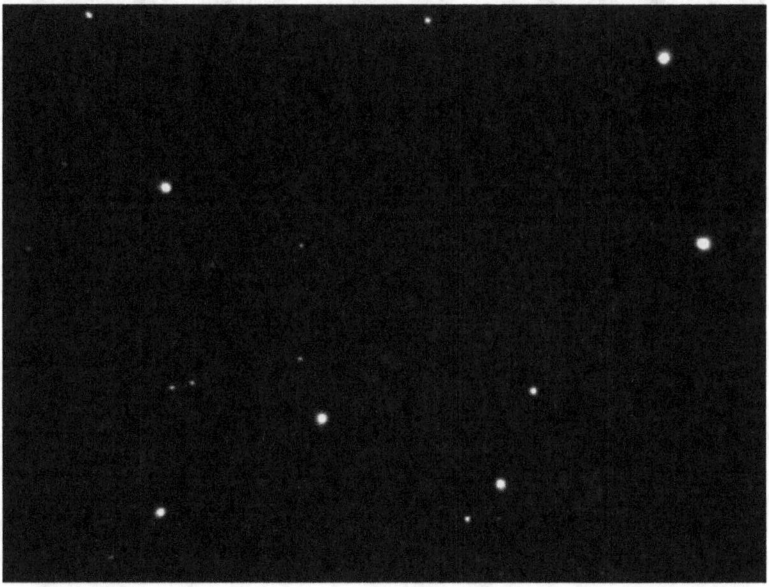

NGC 1981 11″ SCT

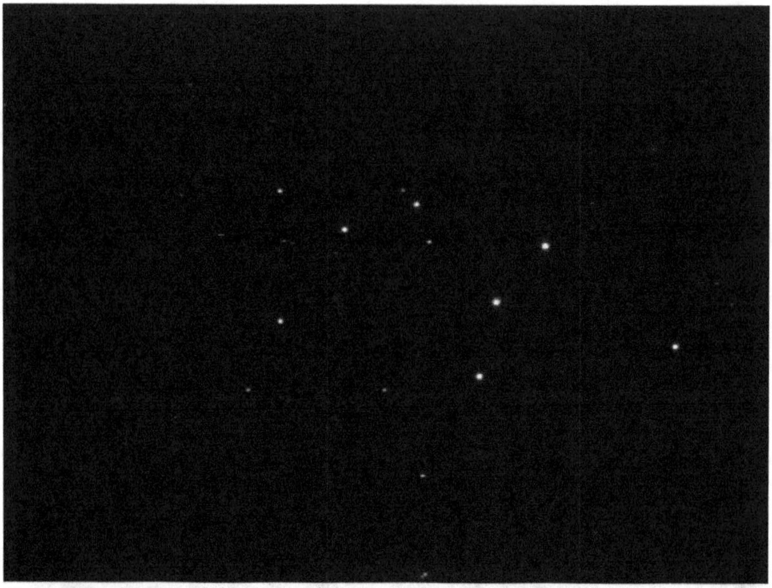

NGC 1981 72mm Refractor

NGC 2022

This planetary nebula is located between two stars, one mag11.6 the other mag12.1, and responds well to high magnifications and UHC or OIII filters.

NGC 2022 11″ SCT

NGC 2022 72mm Refractor

NGC 2024

This emission nebula is better known as the Flame, due to the trisecting dark lanes. Using higher powers and/or an UHC filter helps to eliminate the glare of Zeta Orionis, as it is out of the field of view. Nearby is the famous dark Horsehead Nebula (IC 434), although this object is extremely difficult and requires an H-Beta filter and very large apertures to be seen visually.

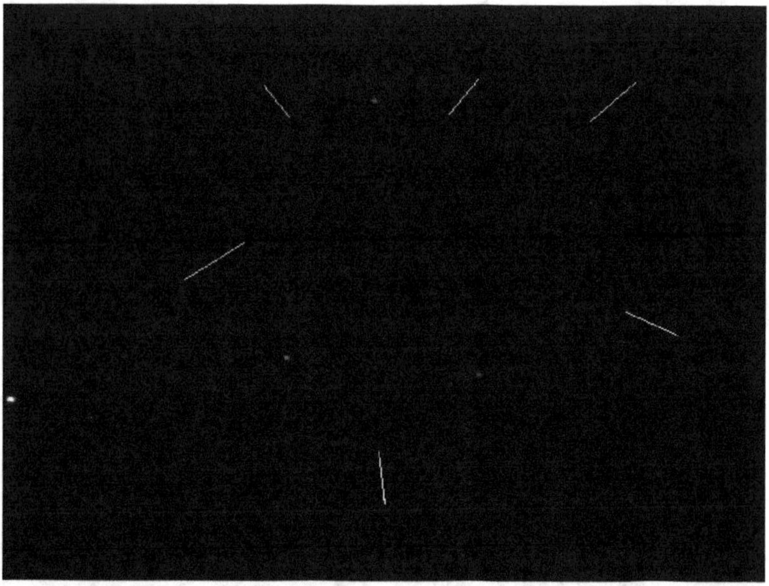

NGC 2024 11″ SCT

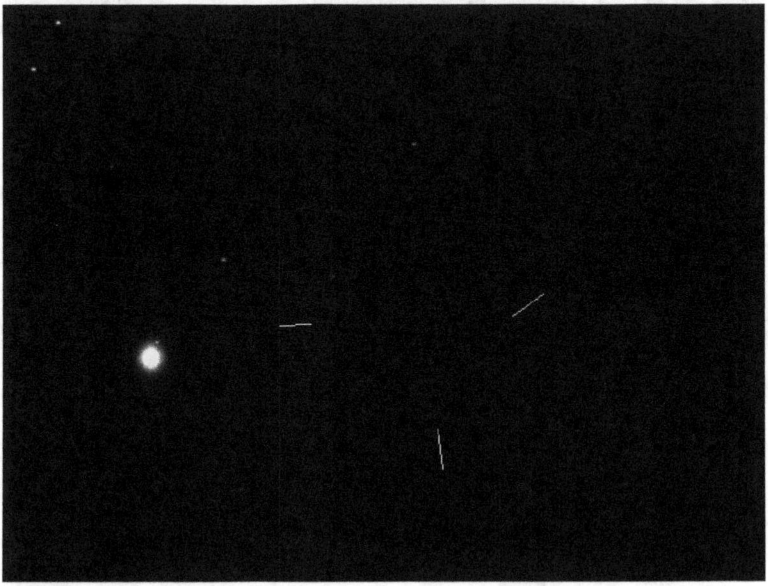

NGC 2024 72mm Refractor

NGC 2194

This open cluster contains almost 200 stars with a moderate range of brightness.

Trumpler Classification: II 2 r

NGC 2194 11″ SCT

NGC 2194 72mm Refractor

Pegasus

Objects	Type	Position	Size	SB*
M 15	GC	21h29′58.3″ +12 10 01	18.0′	
NGC 7331	Galaxy	22h37′04.9″ +34 24 49	10.5′×3.5′	13.2

M 15

This irregularly shaped globular cluster contains the planetary nebula Pease 1, or PK 65-27.1. This planetary is very small (1″), so you will need a very accurate finder chart to identify it! An OIII filter will assist you in your quest.

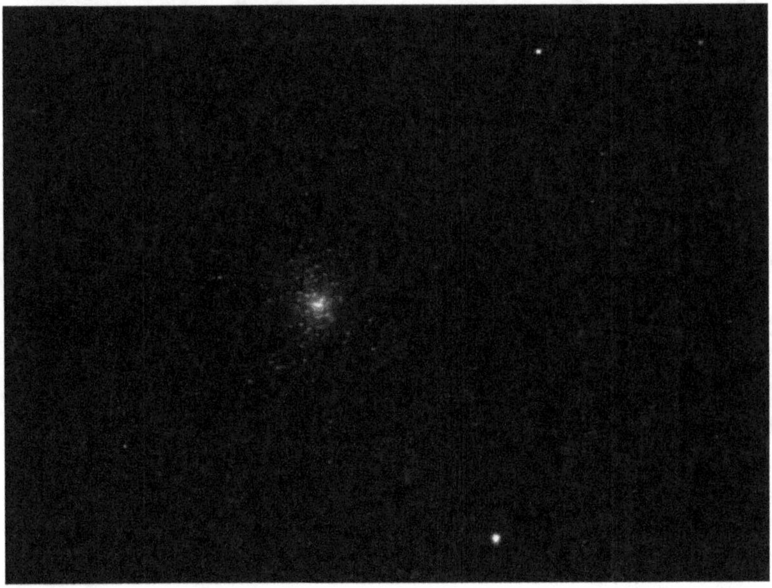

M 15 11″ SCT

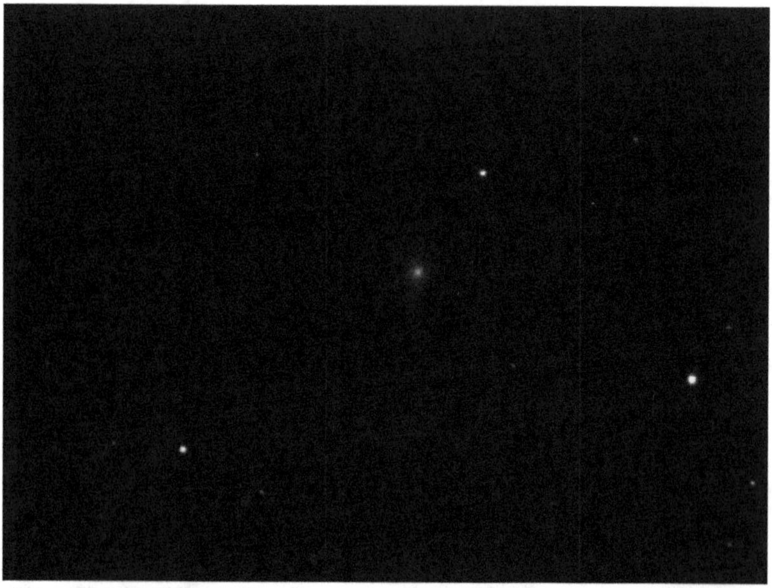

M 15 72mm Refractor

NGC 7331

This is a SAb galaxy presented to us almost edge-on. There are several dark lanes and some galaxies in the field of view. How many friends can you see?

NGC 7331 11″ SCT

NGC 7331 72mm Refractor

Perseus

Objects	Type	Position	Size	SB*
M 34	OC	02h42′08.0″ +42 45 00	25′	
M 76	PN	01h42′19.9″ +51 34 35	187″	10.1
NGC 869 &	OC	02h19′04.0″ +57 08 00	18′	
NGC 884		02h22′05.0″ +57 08 00	18′	
NGC 1023	Galaxy	02h40′24.3″ +39 03 46	7.4′×2.5′	12.4
NGC 1342	OC	03h31′40.0″ +37 23 00	17′	
NGC 1491	EN	04h03′24.0″ +51 19 00	25′×25′	
NGC 1528	OC	04h15′19.0″ +51 13 00	18′	
NGC 1582	OC	04h31′39.0″ +43 45 00	24′	

M 34

This large open cluster contains approximately 100 stars with a large range of brightness. Due to its size, it requires wide fields of view (binoculars?) and low powers.

Trumpler Classification: II 3 m/r

M 34 72mm Refractor

M 76

This planetary nebula is better known as the Little Dumbbell, due to the two 'lobes' visible either side of the core. It has a light green color. An UHC or an OIII filter will help to distinguish this object from the field in small apertures under suburban skies.

M 76 11" SCT

M 76 72mm Refractor

NGC 869 and 884

These two open clusters make up the famous Perseus Double Cluster. Both have a large range of brightness and are concentrated towards the center. In fact, each cluster contains over 300 stars! This is a truly magnificent spectacle and needs a wide field of view to appreciate it!

Trumpler Classification: Both I 3 r

NGC 869 and 884 72mm Refractor

NGC 1023

This is an oval-shaped SB0 galaxy. Can you see its friend, NGC 1023A, only 2.4′ away?

NGC 1023 11″ SCT

NGC 1023 72mm Refractor

NGC 1342

This open cluster contains almost 100 stars with a moderate range of brightness. How many member stars can you see? If it's 50+ you are doing well!

Trumpler Classification: III 2 m

NGC 1342 11″ SCT

NGC 1342 72mm Refractor

NGC 1491

This visually faint emission nebula is made of two 'wisps,' forming a narrow 'V' shape around a mag11 star. An UHC filter will help with the detail. Using averted vision will increase the size of this nebula.

NGC 1491 11″ SCT

NGC 1491 72mm Refractor

NGC 1528

This open cluster is slightly concentrated towards the center and contains approximately 100 stars. Small apertures should reveal about 30–40 member stars. There are two other open clusters nearby, NGCs 1513 and 1545.

NGC 1528 72mm Refractor

Trumpler Classification: II 2 m

NGC 1582

This loose and scattered open cluster contains about 20 stars with a moderate range of brightness. It is large and requires low powers and wide fields of view.

Trumpler Classification: IV 2 p

NGC 1582 72mm Refractor

Pisces

Objects	Type	Position	Size	SB*
M 74	Galaxy	01h36′41.1″ +15 47 16	10.5′×9.5′	14.2

M 74

This is a Sac galaxy with several knots in the halo. Can you see them? Can you see the spiral arms?

M 74 11″ SCT

M 74 72mm Refractor

Puppis

Objects	Type	Position	Size	SB*
M 46	OC (+PN)	07h41′47.0″ −14 49 00	20′	
M 47	OC	07h36′36.0″ −14 29 00	25′	
NGC 2440	PN	07h41′55.4″ −18 12 31	79″	
NGC 2539	OC	08h10′38.0″ −12 49 00	15′	

M 46

This large open cluster contains over 180 stars with a moderate range of brightness. Note the presence of NGC 2438, a small planetary nebula (angular size = 76″) on the northeastern boundary of the cluster (although it itself is not part of the cluster). An OIII filter will make this nebula more prominent.

M 46 11″ SCT

M 46 72mm Refractor

Trumpler Classification: II 2 r

M 47

This open cluster contains about 100 stars with a large range of brightness and is just under 2° away from M 46. M 47 is of a similar apparent size to its friend and requires larger fields of view.

Trumpler Classification: I 3 m/r

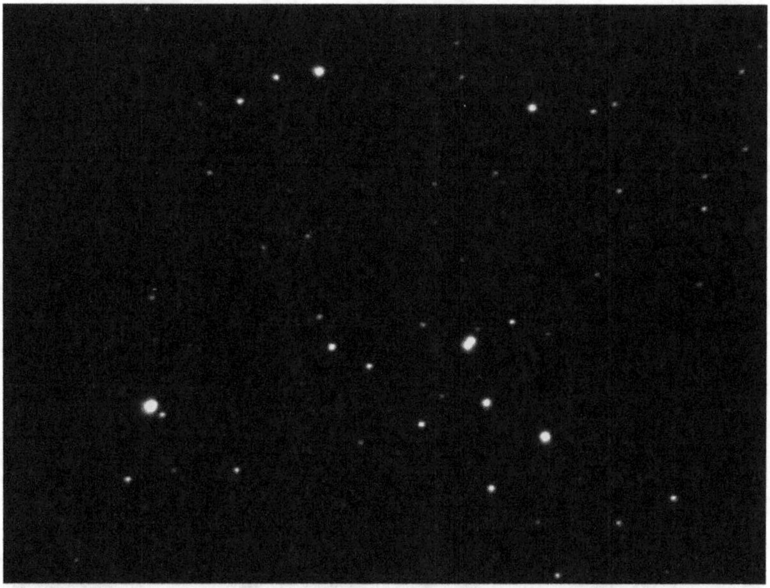

M 47 11″ SCT

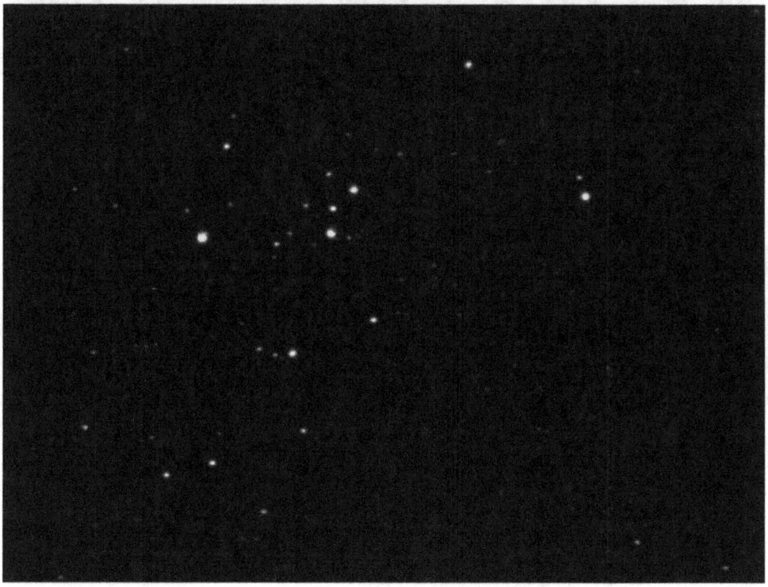

M 47 72mm Refractor

NGC 2440

This planetary nebula has a mag8.4 star only 3′ away. An UHC or an OIII filter will increase contrast with the background. Use higher magnification. Can you see the bipolar structure (just about visible in the 11″ SCT picture below) and the faint extensions?

NGC 2440 11″ SCT

NGC 2440 72mm Refractor

NGC 2539

This open cluster contains almost 60 stars with a moderate brightness range. It is not well detached from the surrounding star field.

Trumpler Classification: I II 2 m

NGC 2539 11″ SCT

NGC 2539 72mm Refractor

Sagitta

Objects	Type	Position	Size	SB*
M 71	GC	19h53′46.1″ +18 46 42	7.2′	

M 71

This is one of the fainter globular clusters in the Messier catalog. It is a very loose globular with an unusual pattern at the core – resembling an arrowhead!

M 71 11″ SCT

M 71 72mm Refractor

Sagittarius

Objects	Type	Position	Size	SB*
M 17	EN	18h20′48.0″ −16 11 00	20′×15′	
M 18	OC	18h20′00.0″ −17 06 00	7′	
M 23	OC	17h56′56.0″ −19 01 00	25′	
M 24 & NGC 6603	SC	18h18′28.0″ −18 24 24	4′	
M 25	OC	18h31′45.0″ −19 07 00	26′	
NGC 6445	PN	17h49′14.9″ −20 00 36	44″	11.2
NGC 6716	OC	18h54′34.0″ −19 55 00	10′	
NGC 6818	PN	19h43′57.8″ −14 09 10	46″	9.3

M 17

This bright emission nebula is more commonly known as the Swan or Omega Nebula due to its obvious '2' shape. Can you see the fainter 'wisps' of nebulosity extending around the '2' and out of the field of view? Using an UHC filter is helpful!

M 17 11″ SCT

M 17 72mm Refractor

M 18

There are approximately 50 stars in this open cluster over a large range of brightness, arranged in two roughly parallel lines. There is some nebulosity associated with this cluster, which is invisible from the author's location. Can you see the nebulosity?

Trumpler Classification: II 3 p n

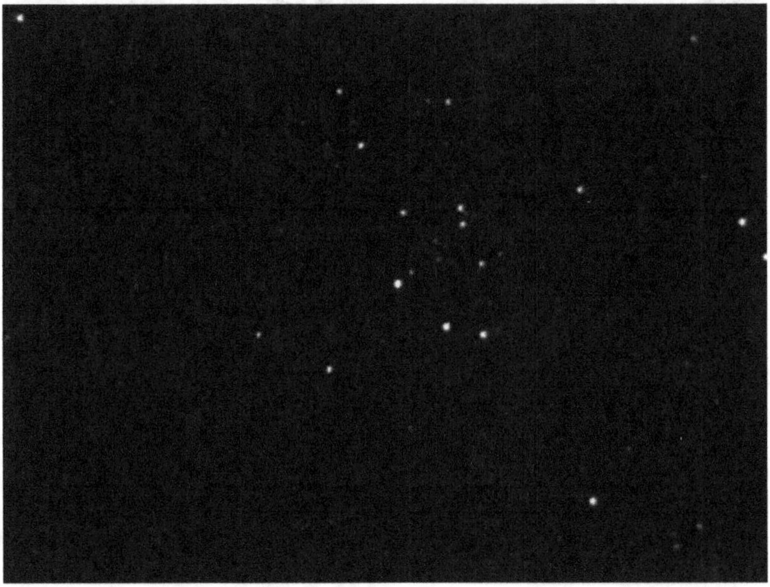

M 18 11" SCT

M 18 72mm Refractor

M 23

This large open cluster contains 150 stars with a moderate range of brightness.

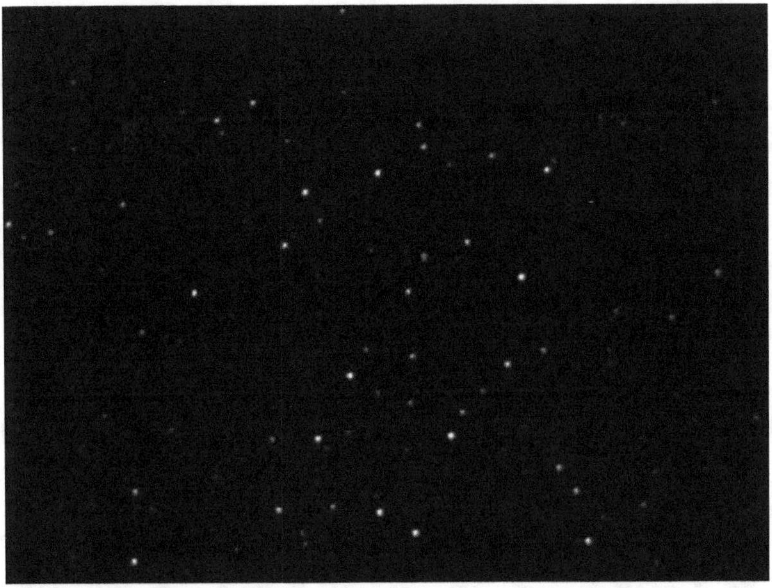

M 23 11" SCT

M 23 72mm Refractor

Trumpler Classification: II 2 r

M 24 and NGC 6603

This object is, in fact, a star cloud whose other designation is IC 4715. There is a small open cluster, NGC 6603, imbedded in it (visible in the center below). The cloud is approximately 2° in size and therefore requires very low powers (binoculars?). However, NGC 6603 is only 4′ in size and well detached from the field.

Trumpler Classification (NGC 6603): I 2 r

M 24 11″ SCT

M 24 72mm Refractor

M 25

This open cluster contains approximately 600 stars, with a large range of brightness.

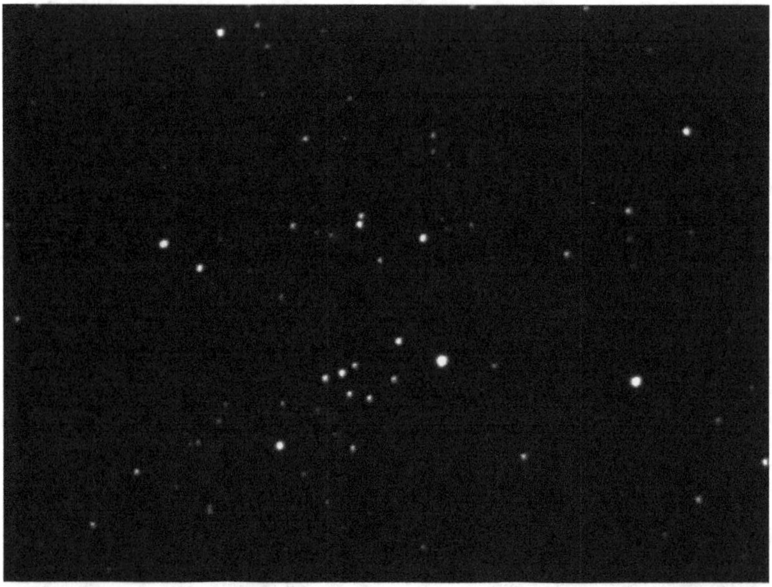

M 25 11″ SCT

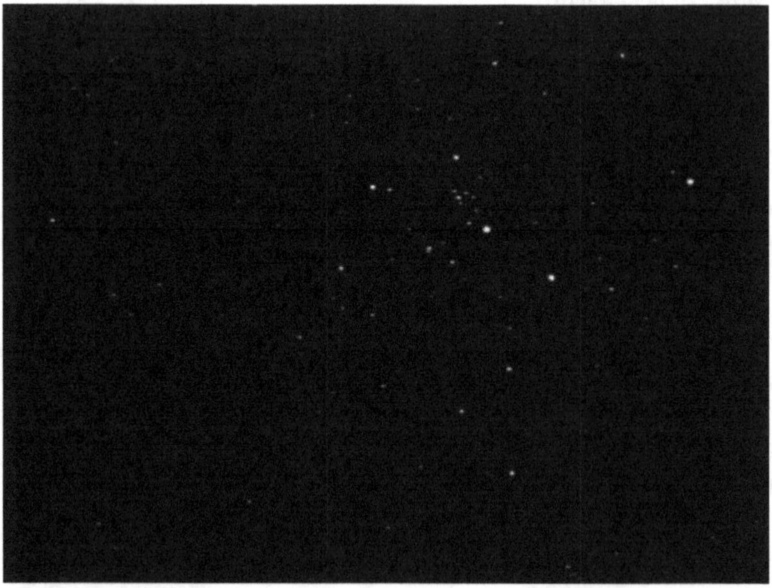

M 25 72mm Refractor

Trumpler Classification: I 3 m/r

NGC 6445

This planetary nebula has a rectangular slight 'dumbbell' shape with a slight darkening towards the center and is very similar in appearance to M 76, the Little Dumbbell, in Perseus. Using an UHC filter will make this object stand out more.

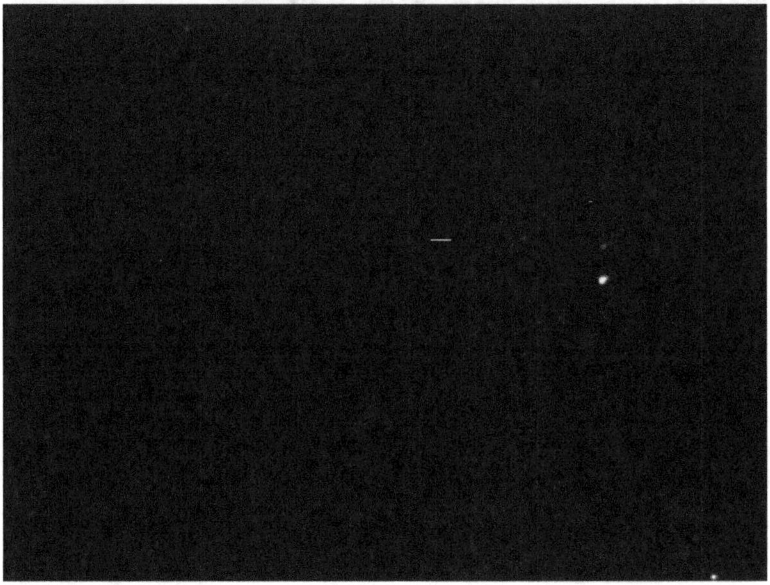

NGC 6445 11″ SCT

NGC 6445 72mm Refractor

NGC 6716

This loose open cluster contains almost 40 stars of similar brightness. It is not well detached from the field.

NGC 6716 11″ SCT

NGC 6716 72mm Refractor

Trumpler Classification: IV 1 p

NGC 6818

This planetary nebula is more commonly known as the Little Gem due to its brightness and elongated shape. Can you see the mag13.5 stars on the eastern and western edges? Can you make out any texture within the nebula?

NGC 6818 11″ SCT

NGC 6818 72mm Refractor

Scutum

Objects	Type	Position	Size	SB*
M 11	OC	18h51′04.0″ –06 16 00	11′	
M 26	OC	18h45′15.0″ –09 23 00	10′	
NGC 6712	GC	18h53′04.3″ –08 42 22	9.8	

M 11

This open cluster is also known as the Wild Duck cluster and contains almost 700 stars. It is very condensed towards the center and appears much like a globular at low powers.

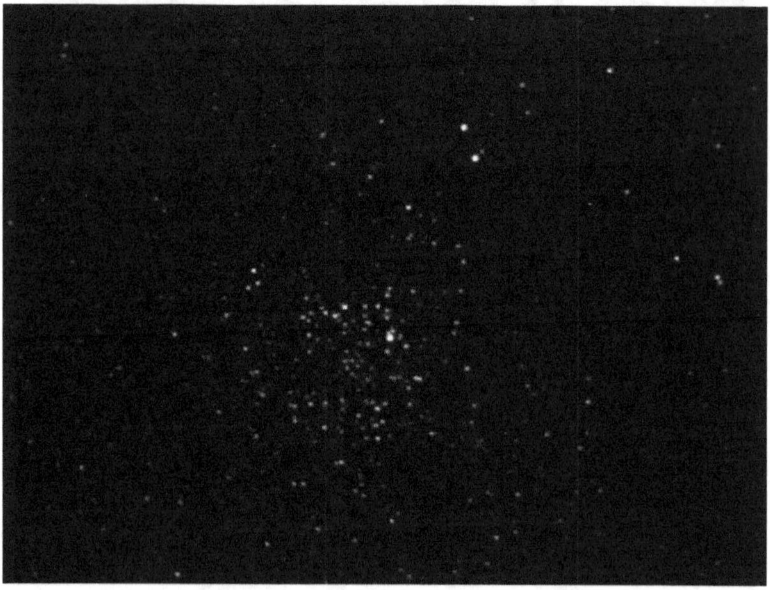

M 11 11″ SCT

M 11 72mm Refractor

Trumpler Classification: I 2 r

M 26

This open cluster is close to M 11 but fainter and contains 120 stars with a large range of brightness. Many of these stars are beyond the reach of amateur equipment, hence the 'm' classification for richness.

Trumpler Classification: II 3 m

M 26 11″ SCT

M 26 72mm Refractor

NGC 6712

This loose globular cluster has an irregular shape and requires larger apertures (probably 10″+) to resolve any stars.

NGC 6712 11″ SCT

NGC 6712 72mm Refractor

Serpens Caput

Objects	Type	Position	Size	SB*
M 5	GC	15h18'33.8" +02 04 58	23'	

M 5

This apparently large globular cluster has a slightly oval shape. Larger apertures should resolve the core nicely. Can you?

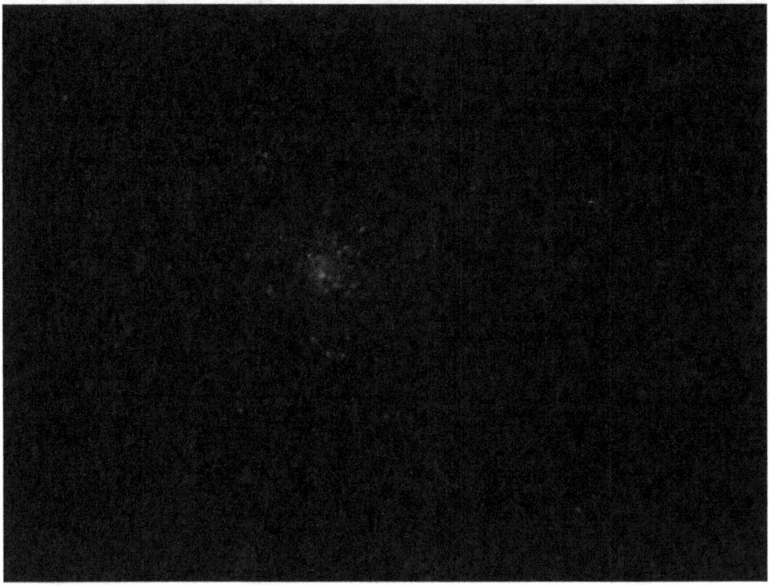

M 5 11" SCT

M 5 72mm Refractor

Serpens Cauda

Objects	Type	Position	Size	SB*
M 16+IC 4703	OC+EN	18h18′45.0″ −13 48 00 18h18′36.0″ −13 58 00	8′35′×28′	
IC 4756	OC	18h38′54.0″ +05 26 00	40′	

M 16+IC 4703

This open cluster contains over 500 stars, about 60 of which are detectable, and is surrounded by the famous emission nebula, the Eagle Nebula (IC 4703). In this area there is an absorption patch that loosely resembles an eagle in flight – not visible in suburbia. How much nebulosity can you make out?

Trumpler Classification: II 3 m n

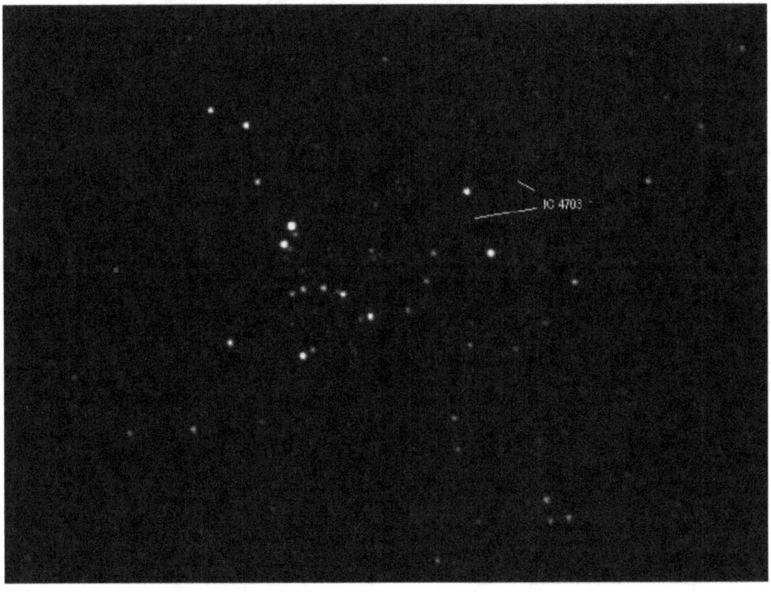

M 16+IC 4703 11″ SCT

M 16 72mm Refractor

IC 4756

This very large open cluster contains over 450 stars but is best viewed at low powers and with wide fields of view (large binoculars?).

Trumpler Classification: II 3 r

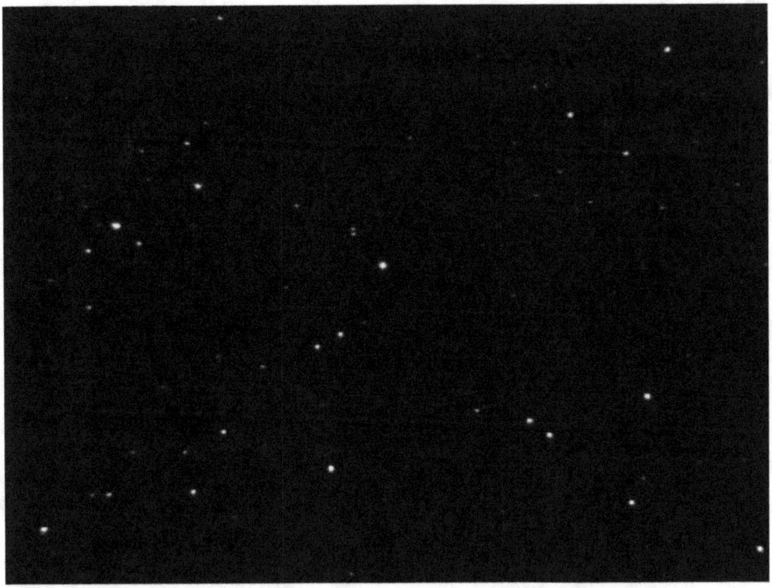

IC 4756 11″ SCT

IC 4756 72mm Refractor

Sextans

Objects	Type	Position	Size	SB*
NGC 3115	Galaxy	10h05′14.1″ −07 43 07	7.2′×2.4′	11.9

NGC 3115

This S0sp galaxy is a good example of an edge-on spindle with a bright central region. The galaxy UGCA 200 is only 5.6′ away.

NGC 3115 11″ SCT

NGC 3115 72mm Refractor

Taurus

Objects	Type	Position	Size	SB*
M 1	SR-E	05h34′30.0″ +22 01 00	6′×4′	
M 45	OC	03h47′29.0″ +24 06 00	120′	
NGC 1514	PN	04h09′16.9″ +30 46 34	132″	10.9
NGC 1647	OC	04h45′42.2″ +19 07 00	40′	
NGC 1746	OC	05h03′36.0″ +23 49 00	40′	
NGC 1807	OC	05h10′46.0″ +16 31 00	12′	
NGC 1817	OC	05′12′27.0″ +16 41 00	20′	

M 1

This famous supernova remnant is commonly known as the Crab Nebula and visually resembles a faint parallelogram. Can you make out any of the inner detail (the faint 'tendrils' of the Crab)? Use an UHC filter for this one.

M 1 11″ SCT

M 1 72mm Refractor

M 45

This open cluster is commonly known as the Pleiades. It contains 100 stars with a large range of brightness. This is best viewed using low powers. Can you see the nebulosity, particularly around Merope?

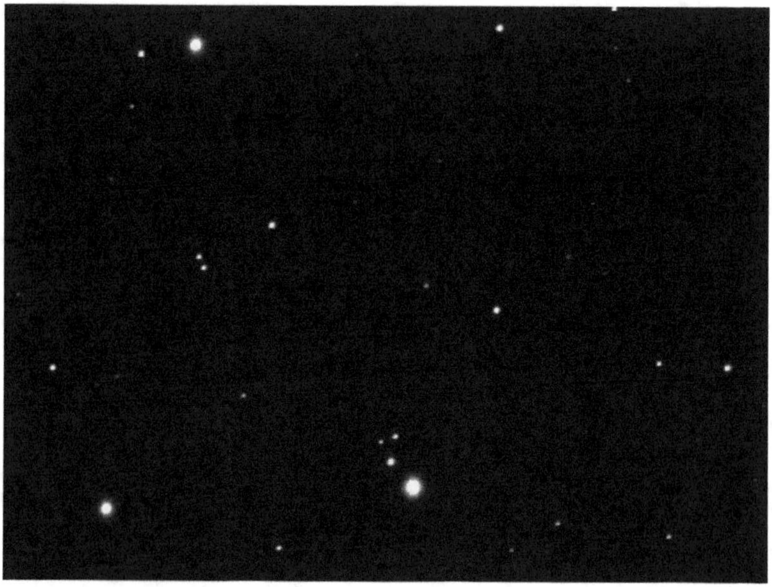

M 45 72mm Refractor

Trumpler Classification: I 3 r n

NGC 1514

This visually faint planetary nebula has a strong dark patch close to the bright (mag9.4) central star. Use an OIII or an UHC and see if you can find the extra detail within. Note that in the unfiltered pictures below the nebula is barely visible in either instrument.

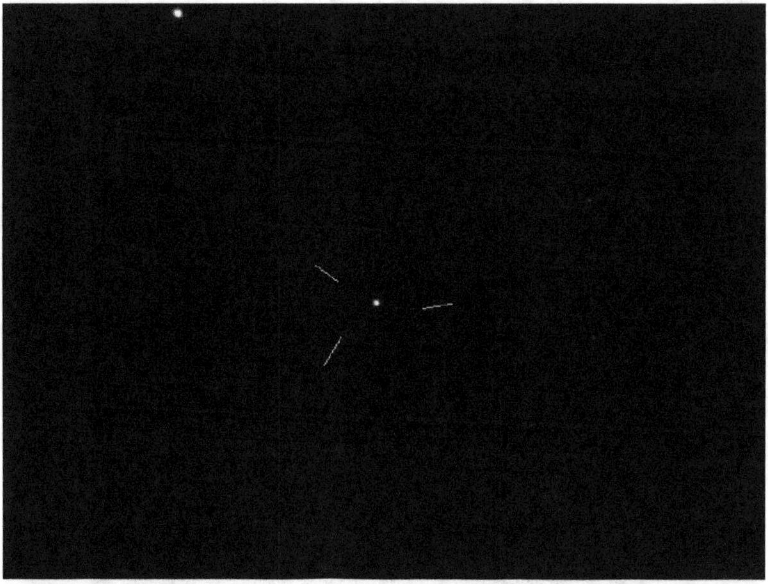

NGC 1514 11″ SCT

NGC 1514 72mm Refractor

NGC 1647

This large open cluster contains 200 stars with a moderate range of brightness. It appears very loose despite its 'II' Trumpler classification.

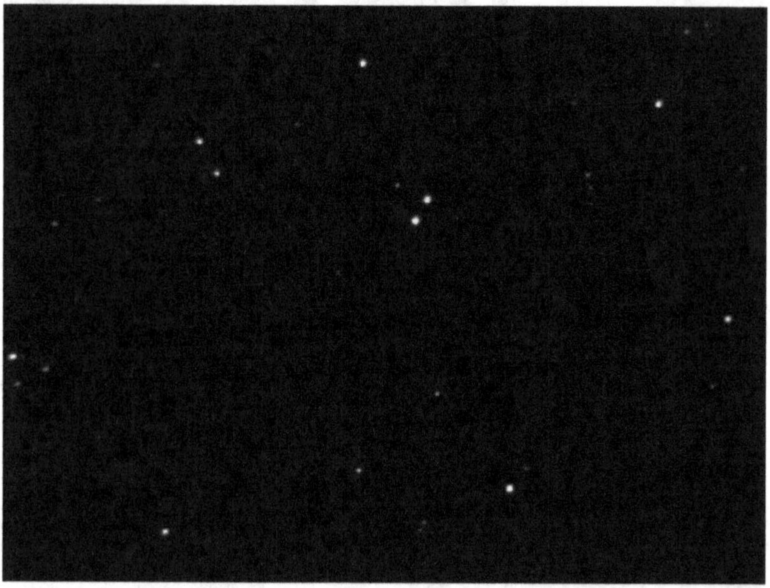

NGC 1647 11″ SCT

NGC 1647 72mm Refractor

Trumpler Classification: II 2 r

NGC 1746

This open cluster contains only 20 loosely scattered stars with a moderate brightness range. It requires low powers (binoculars?) and wide fields of view.

Trumpler Classification: III 2 p

NGC 1746 11″ SCT

NGC 1746 72mm Refractor

NGC 1807

There are nearly 40 stars with a moderate range of brightness in this open cluster. It can be seen in the same field of view as NGC 1817 at low power.

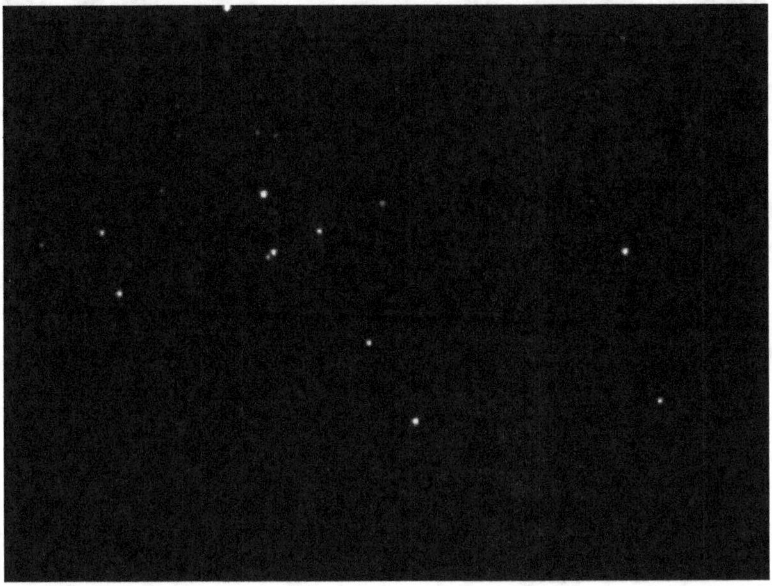

NGC 1807 11″ SCT

NGC 1807 72mm Refractor

Trumpler Classification: II 2 p

NGC 1817

This open cluster contains almost 300 stars with a moderate range of brightness. It can be found in the same low power field of view as NGC 1807.

Trumpler Classification: IV 2 r

NGC 1817 11″ SCT

NGC 1817 72mm Refractor

Triangulum

Objects	Type	Position	Size	SB*
M 33	Galaxy	01h33′51.3″ +30 39 54	70.8′×41.7′	14.2

M 33

This SAcd galaxy is also known as the Pinwheel Galaxy. It has a low surface brightness and is difficult in small apertures. Can you see the HII regions? A lot of these have their own NGC and IC designations (see NGC 604 below). The diffuse core of M 33 is to the left in the 11″ SCT picture below and in the center of the 72mm refractor view.

M 33 11″ SCT

M 33 72mm Refractor

Ursa Major

Objects	Type	Position	Size	SB*
M 40	DS	12h21′54.0″ +58 05 00		
M 81	Galaxy	09h55′38.5″ +69 03 40	26.9′×14.1′	13.2
M 82	Galaxy	09h55′52.2″ +69 40 48	11.2′×4.3′	12.5
M 97	PN	11h14′47.8″ +55 01 09	170″	9.9
M 101	Galaxy	14h03′12.7″ +54 21 12	28.8′×26.9′	14.9
M 108	Galaxy	11h11′29.6″ +55 40 22	8.7′×2.2′	13.1
M 109	Galaxy	11h57′35.7″ +53 22 28	7.6′×4.6′	13.5
NGC 2841	Galaxy	09h22′02.4″ +50 58 39	8.1′×3.5′	12.7
NGC 3079	Galaxy	10h01′57.9″ +55 40 36	7.9′×1.4′	13.3
NGC 3184	Galaxy	10h18′17.0″ +41 25 20	7.4′×6.9′	13.9
NGC 3877	Galaxy	11h46′07.3″ +47 29 39	5.5′×1.3′	13.0
NGC 3941	Galaxy	11h52′55.1″ +36 59 05	3.5′×2.3′	12.5
NGC 4026	Galaxy	11h59′25.4″ +50 57 41	5.2′×1.3′	12.7
NGC 4088	Galaxy	12h05′33.5″ +50 32 14	5.8′×2.2′	13.2
NGC 4157	Galaxy	12h11′04.2″ +50 29 06	6.8′×1.1′	13.4
NGC 4605	Galaxy	12h40′01.0″ +61 36 33	5.8′×2.2′	12.9

M 40

This is actually a double star, also known as Winnecke 4. Both stars are separated by 50″.

M 40 11″ SCT

M 40 72mm Refractor

M 81

This Saab galaxy is commonly known as Bode's Nebula and is close to M 82 – in the same low power field of view. Can you see any structure (spiral arms) in the halo? Can you see UGC 5336 only 4.5′ to the east?

M 81 11″ SCT

M 81 72mm Refractor

M 82

This galaxy is classed I0sp and is presented edge-on with very prominent, jagged dust lanes. As mentioned above, it is in the same low power field of view as M 81.

M 82 11″ SCT

M 82 72mm Refractor

M 97

This planetary nebula is better known as the Owl Nebula and is best viewed using an UHC or an OIII filter. It is very faint in the pictures below (use averted vision). Can you see the famous eyes? Note also that M 108 (see later) is only 48′ away.

M 97 11″ SCT

M 97 72mm Refractor

M 101

This SABcd galaxy is also (possibly) M 102, the famous duplicate Messier observation. The core is clearly visible, but the famous spiral arms of this galaxy are very difficult. Can you see them? If so, there are several HII regions and star clusters within them that have their own NGC and IC designations!

M 101 11″ SCT

M 101 72mm Refractor

M 108

This is a SBcd galaxy located very close to M 97 (only 48′ away) and is presented to us edge-on. There is a mag12 star superimposed just west of center. Can you see the knot just west of the core?

M 108 11″ SCT

M 108 72mm Refractor

M 109

This SBbc galaxy contains a brighter core region and a very faint outer halo. Can you see the mottling near the core? If not, use a higher power.

M 109 11″ SCT

M 109 72mm Refractor

NGC 2841

This SAb galaxy is just 4.6′ away from a mag8.5 star and has a small, bright core. Note the dust lane on the eastern side cutting off some of the galaxy from view.

NGC 2841 11″ SCT

NGC 2841 72mm Refractor

NGC 3079

This SBc sp galaxy is presented to us edge-on. Can you see NGC 3073 (near the edge of the field in the picture below)? PGC 28990 is also very close by – can you see it?

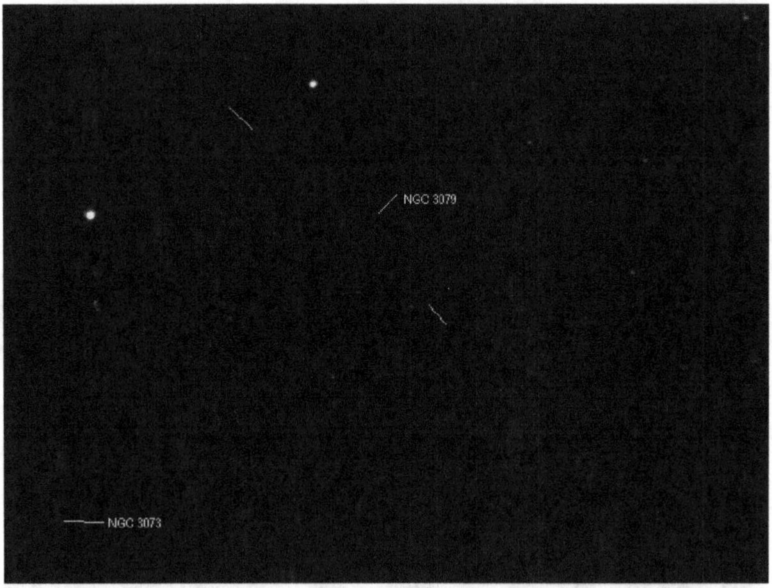

NGC 3079 11″ SCT

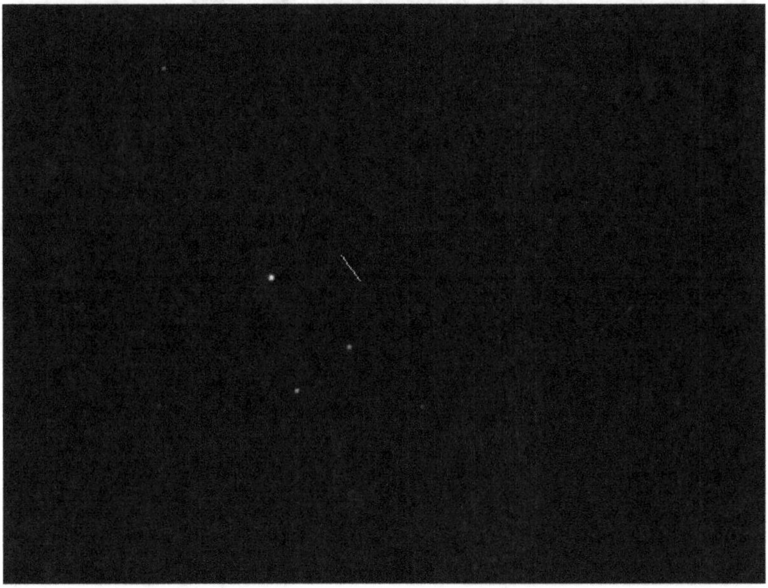

NGC 3079 72mm Refractor

NGC 3184

This SABcd galaxy is presented to us face-on and so has a low surface brightness. Can you make out any structure in the faint halo (spiral arms)? Note the superimposed mag11.5 star on the northern edge.

NGC 3184 11″ SCT

NGC 3184 72mm Refractor

NGC 3877

This Sac galaxy has a mag9.9 star just 3.7′ away and is presented to us edge-on.

NGC 3877 11″ SCT

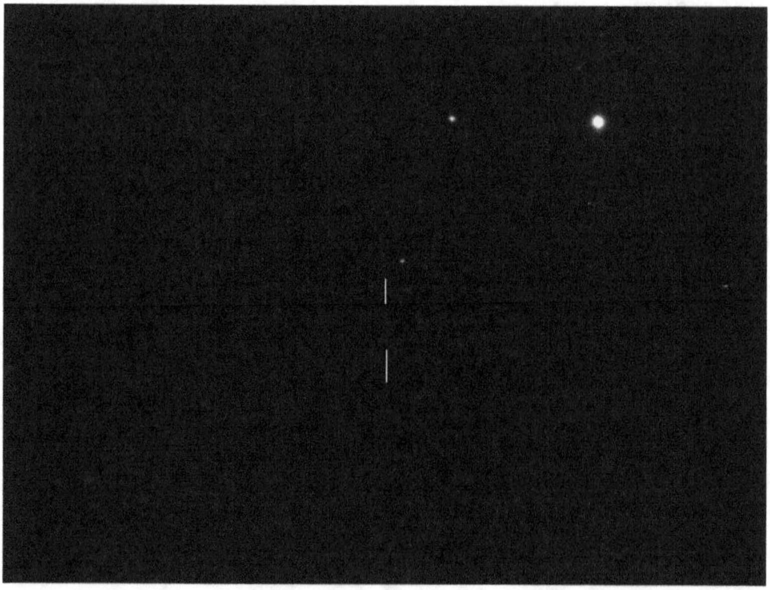

NGC 3877 72mm Refractor

NGC 3941

This is a SB0 galaxy with a very bright central region. How much of the halo can you discern?

NGC 3941 11″ SCT

NGC 3941 72mm Refractor

NGC 4026

This is a bright S0sp galaxy presented to us edge-on. Note the bright oval core. Can you spot UGC 6956, also known as PGC 37682, almost 10′ away (just out of the field of view on the left in the 11″ SCT picture below)?

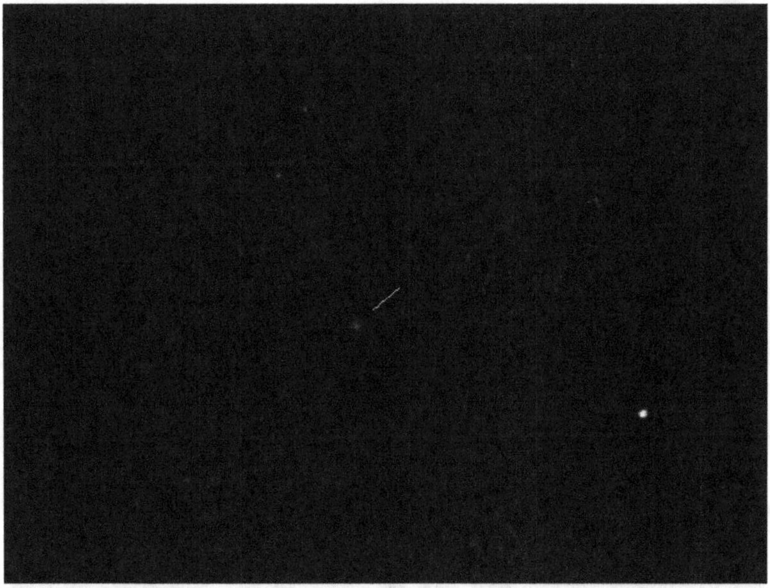

NGC 4026 11" SCT

NGC 4026 72mm Refractor

NGC 4088

This is a SABbc galaxy with an extended, brighter central region and is located between NGCs 4026 and 4157. Can you see any structure (just visible in the picture below)? Two dark lanes give this galaxy a devil's fork or cactus-like appearance. This object was invisible in the author's 72mm refractor.

NGC 4088 11″ SCT

NGC 4157

This SABb galaxy is presented to us edge-on and is situated between a mag10.4 star and a mag8.1 star and close to NGCs 4088 and 4026.

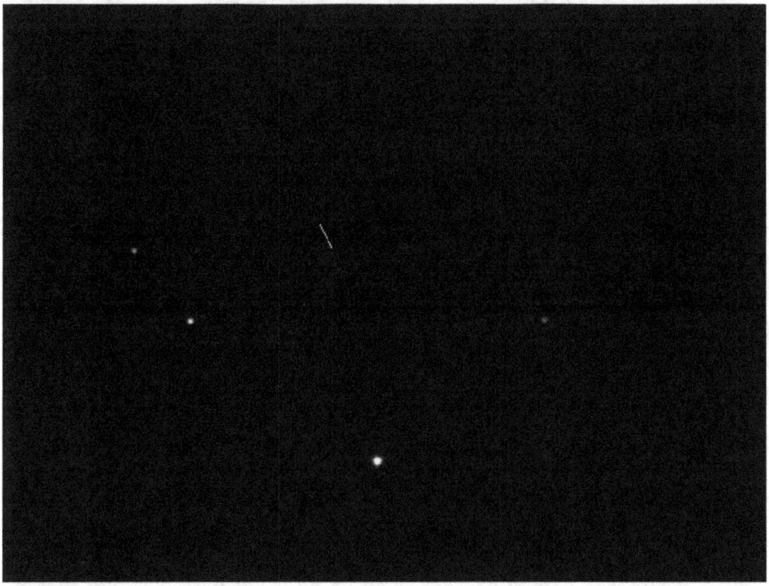

NGC 4157 11″ SCT

NGC 4157 72mm Refractor

NGC 4605

This is a SBc pec galaxy presented to us edge-on with a bright, patchy center and an overall 'bird in flight' shape. Can you make out any knots in the southeast?

NGC 4605 11″ SCT

NGC 4605 72mm Refractor

Ursa Minor

Objects	Type	Position	Size	SB*
NGC 6217	Galaxy	16h32′38.6″ +78 11 55	3.0′×2.5′	13.2

NGC 6217

This SBbc galaxy has a long bar and a dark patch just northeast of the nucleus. Can you make out any of these details? Note that only a hint of the bar is visible around the core in the pictures below.

NGC 6217 11″ SCT

NGC 6217 72mm Refractor

Virgo

Objects	Type	Position	Size	SB
M 49	Gx	12h29′46.7″ +08 00 00	10.2×8.3	13.2
M 58	Gx	12h37′43.7″ +11 49 06	6.0×4.8	13.1
M 59	Gx	12h42′02.2″ +11 38 50	5.4×3.7	13.0
M 60	Gx	12h43′39.8″ +11 33 11	7.6×6.2	13.1
M 61	Gx	12h21′54.9″ +04 28 22	6.5×5.9	13.1
M 84	Gx	12h25′03.6″ +12 53 13	6.5×5.6	13.2
M 86	Gx	12h26′11.5″ +12 56 47	8.9×5.8	13.3
M 87	Gx	12h30′49.4″ +12 23 26	8.3×6.6	13.0
M 89	Gx	12h35′39.9″ +12 33 22	3.5×3.5	12.7
M 90	Gx	12h36′50.0″ +13 09 50	9.5×4.4	13.3
M 104	Gx	12h39′59.3″ −11 37 21	8.6×4.2	12.0
NGC 4216	Gx	12h15′54.3″ +13 09 02	8.1×1.8	12.8
NGC 4388	Gx	12h25′47.4″ +12 39 42	5.6×1.3	13.0
NGC 4438	Gx	12h27′45.3″ +13 00 30	8.5×3.2	13.6
NGC 4517	Gx	12h32′45.8″ +00 06 52	10.5×1.5	13.2
NGC 4526	Gx	12h34′02.5″ +07 42 03	7.2×2.4	12.7
NGC 4535	Gx	12h34′19.6″ +08 11 47	7.1×5.0	13.7
NGC 4567	Gx	12h36′32.4″ +11 15 34	3.0×2.0	13.1
NGC 4568	Gx	12h36′33.9″ +11 14 07	4.6×2.0	13.1
NGC 4699	Gx	12h49′02.2″ – 08 39 50	4.0×2.8	10.9
NGC 4762	Gx	12h52′55.3″ +11 13 39	8.7×1.7	13.1
NGC 5746	Gx	14h44′55.8 +01 57 22	7.4×1.3	12.6

M 49

This is a reasonably bright E2 galaxy with a very bright core. Can you see the halo extend beyond the 12th mag star at the eastern edge?

M 49 11" SCT

M 49 72mm Refractor

M 58

At first glance, this object appears to resemble M 49. However, this is a SBb galaxy. Can you make out the slight elongation of the halo?

M 58 11" SCT

M 58 72mm refractor

M 59

This E5 galaxy has a bright core and a stellar nucleus. Does the halo reach the 13th mag star at the northern end?

M 59 11″ SCT

M 59 72mm refractor

M 60

This is an E2 galaxy, with no discernable structure and a bright nucleus. NGC 4647 is just visible nearby. Can you see it?

M 60 11″ SCT

M 60 72mm refractor

M 61

This is a SBbc galaxy. As can be seen below, it has a stellar core. Can you see the arms? Two or three are visible with averted vision, especially in larger telescopes.

M 61 11" SCT

M 61 72mm Refractor

M 84 and M 86

Both of these galaxies (E1 and E3 respectively) are easily visible in the same field, together with their friend NGC 4387. The two Messier objects appear very similar, but can you make out the slightly elongated halo of M 86 compared to M 84?

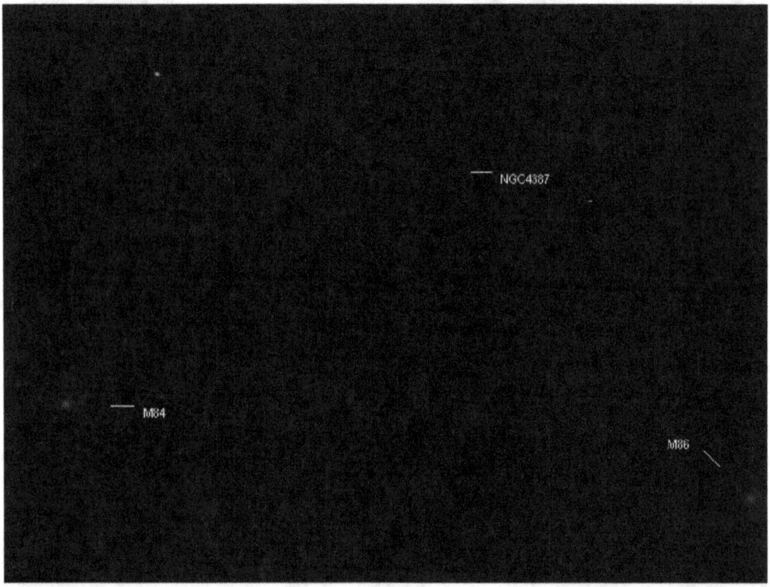

M 84 and 86 11″ SCT

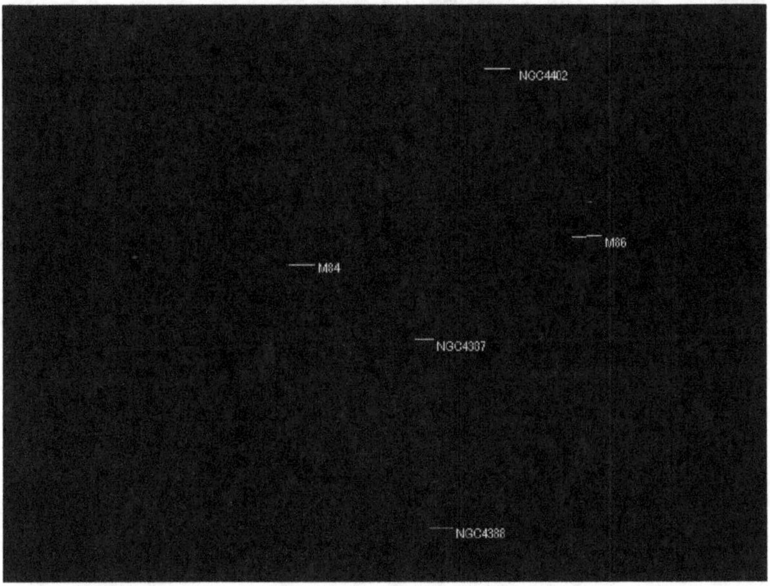

M 84 and 86 and NGCs 4387, 4388, and 4402 (see later) 72mm Refractor

M 87

This large galaxy is classed as E pec. Note the NGC galaxies in the same field and the steady increase in brightness towards the core of M 87. The challenge here is to see the famous 'jet.' It should be just visible in very large apertures (~20").

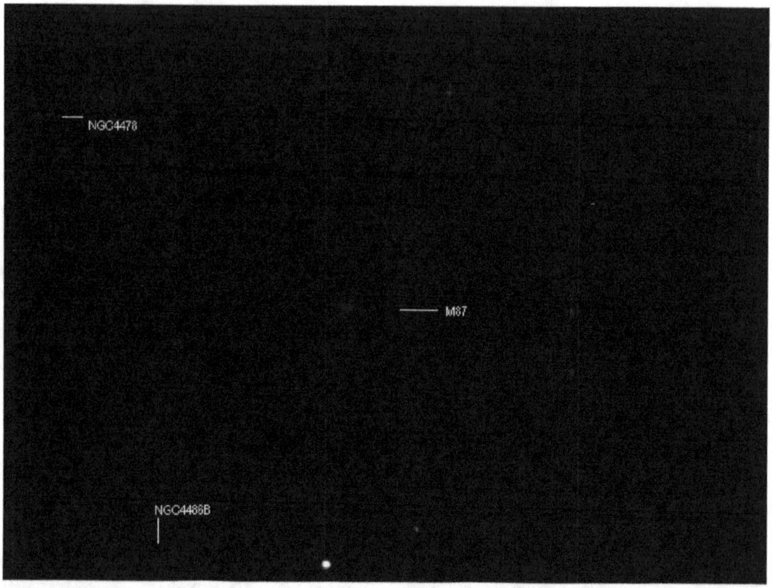

M 87 and friends 11″ SCT

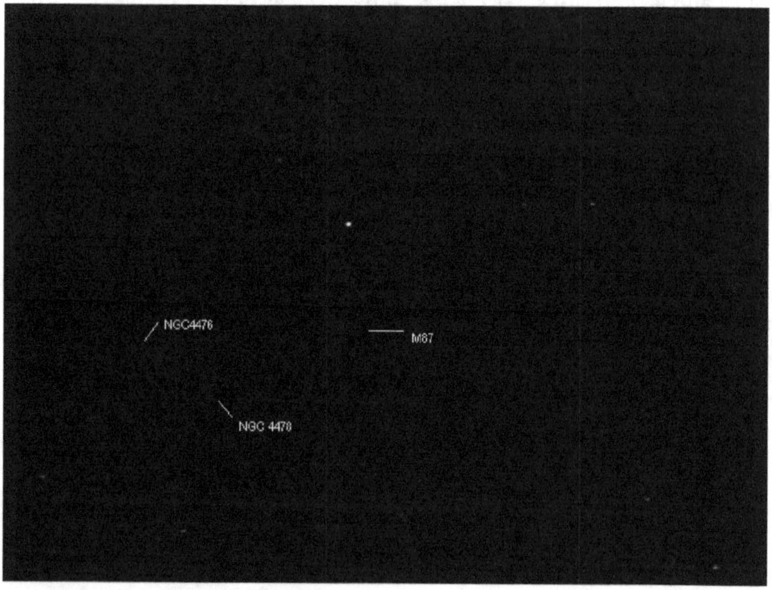

M 87 and friends 72mm Refractor

M 89

This E1 galaxy has a small, bright core and an almost stellar nucleus. There is not much extra detail to be gleaned from this one even in larger apertures. Can you see the faint star on the eastern edge?

M 89 11″ SCT

M 89 72mm Refractor

M 90

This SBab galaxy appears to have an almost stellar nucleus. Can you make out the elongated halo (just visible in the 11″ SCT image below)?

M 90 11″ SCT

M 90 72mm Refractor

M 104

This famous Sa galaxy bears the name the Sombrero Galaxy, due to its distinctive shape. The bisecting dust lane and bright northern sections are easy to see. How much of the southern section is visible to you?

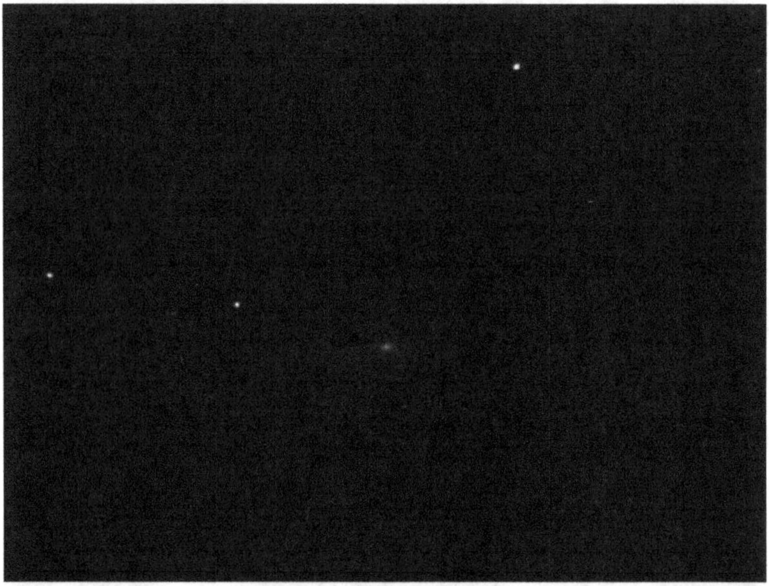

M 104 11″ SCT

M 104 72mm Refractor

NGC 4216

This SAb galaxy is presented edge-on and has a dust lane that runs along the eastern side of it (next to the elongated core). Can you make it out? There are two other galaxies in the field on the limits of visibility – NGC 4206 and 4222 (use averted vision!).

NGC 4216 11″ SCT

NGC 4216 72mm Refractor

NGC 4388

This SAb galaxy is very close to – within 1° of – M 84 and M 86. Can you make out the extent of the elongated halo? Also, in the same field is the fainter NGC 4387.

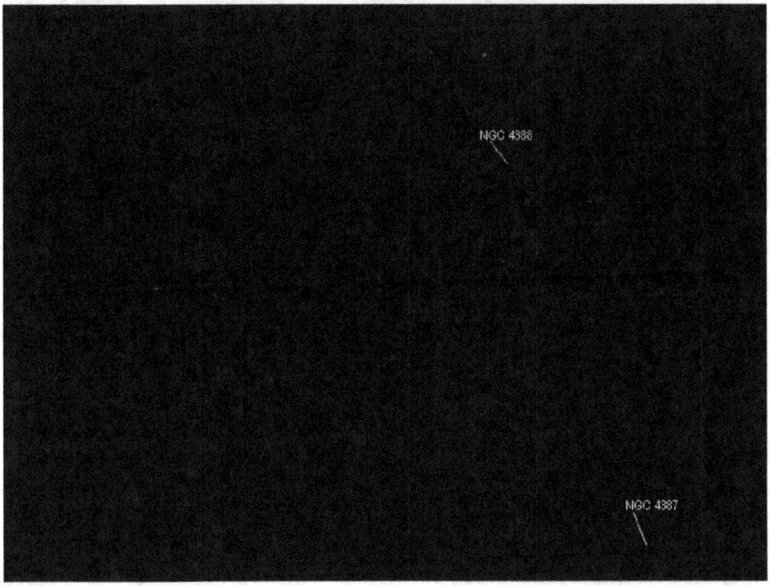

NGC 4388 11″ SCT

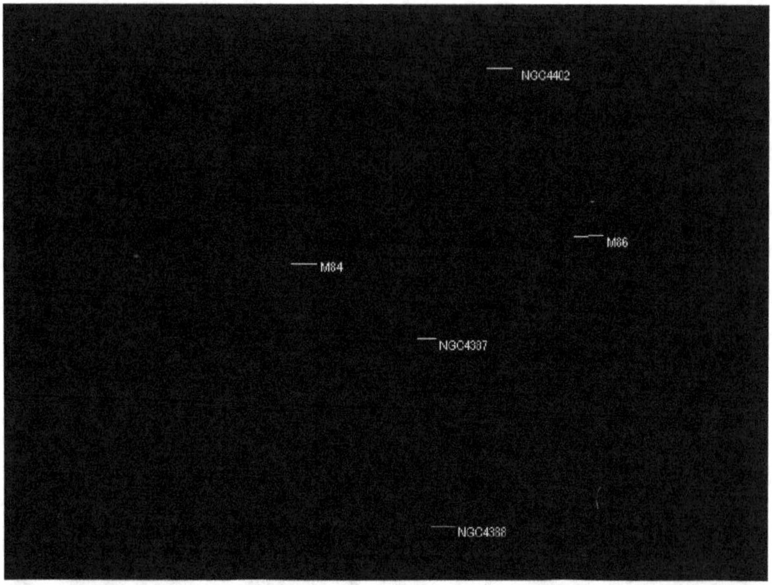

NGC 4388 and friends (see earlier M 84 and M 86) 72mm Refractor

NGC 4438

This galaxy has the classification of SA0/a pec and a lower surface brightness than its friend NGC 4435 nearby. Both of these galaxies are very close to M 84 and M 86 and are sometimes referred to as the Eyes. How much of the length of 4438 can you make out?

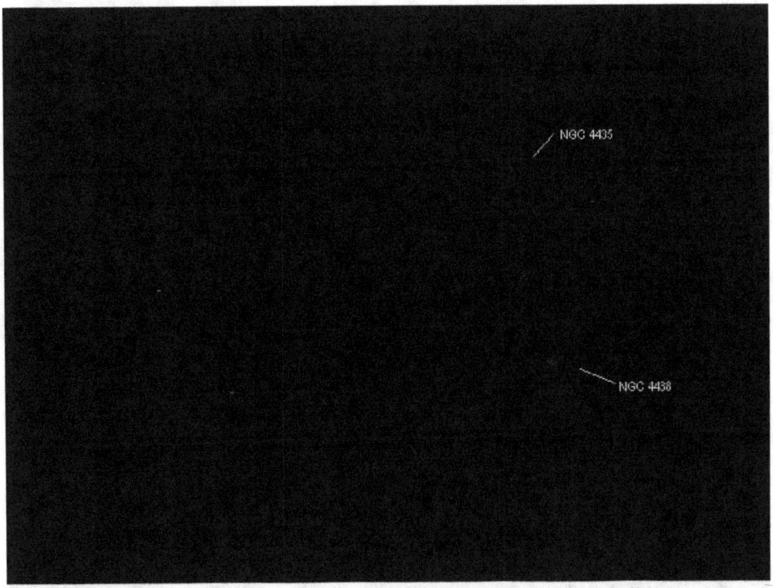

NGC 4438 11″ SCT

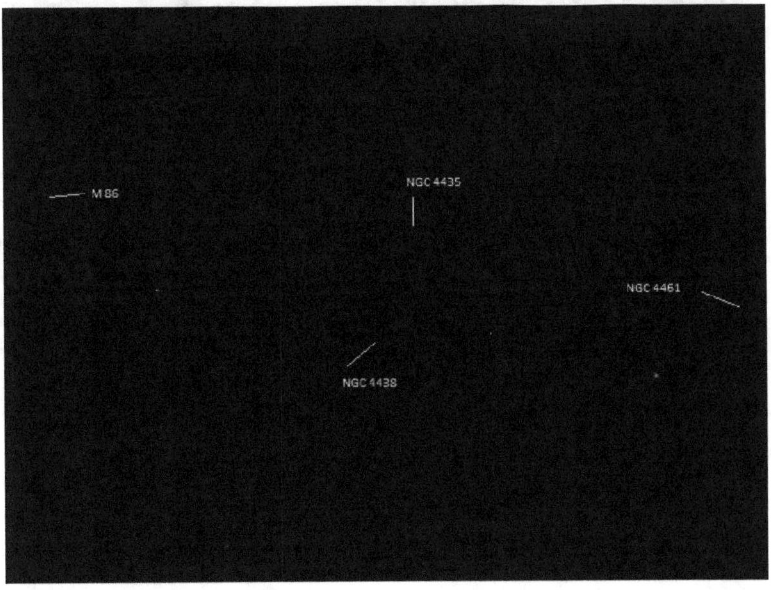

NGC 4438 72mm Refractor

NGC 4526

This SAB galaxy is almost exactly halfway between two bright stars. It is elongated with a bright, almost stellar, nucleus. Can you see the star 1.3′ from the center? How much of the halo can you make out? NGCs 4518 and 4518B are also very close by.

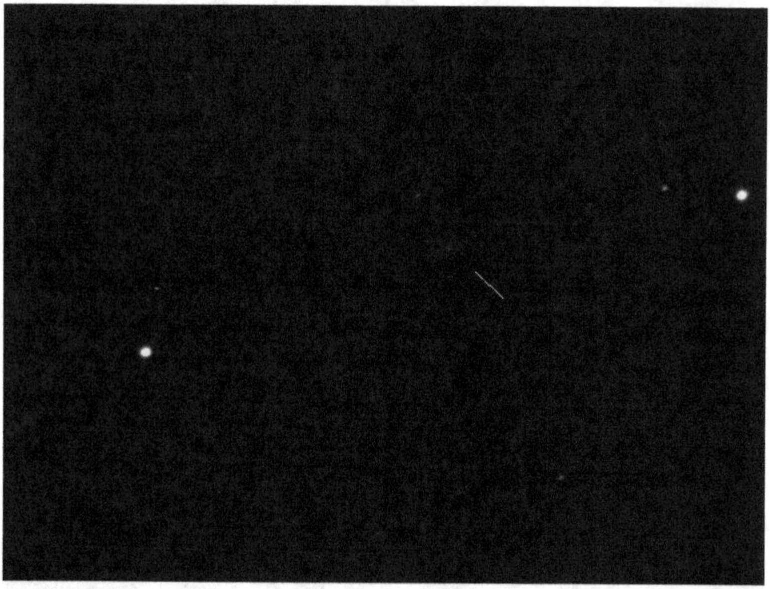

NGC 4526 11″ SCT

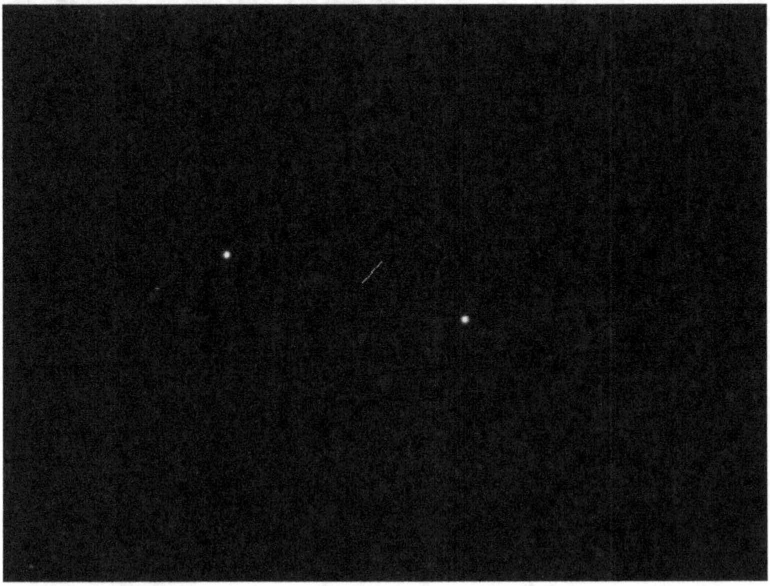

NGC 4526 72mm Refractor

NGC 4535

This SAB galaxy is presented to us face-on. The core appears slightly elongated. The challenge here is to see any the spiral arms. Look for the darkening either side of the core.

NGC 4535 11″ SCT

NGC 4535 72mm Refractor

NGC 4567 and 4568

These two interacting SAbc galaxies are usually referred to as the Siamese Twins. If you have a large aperture telescope, can you make out the dark lanes in NGC 4568?

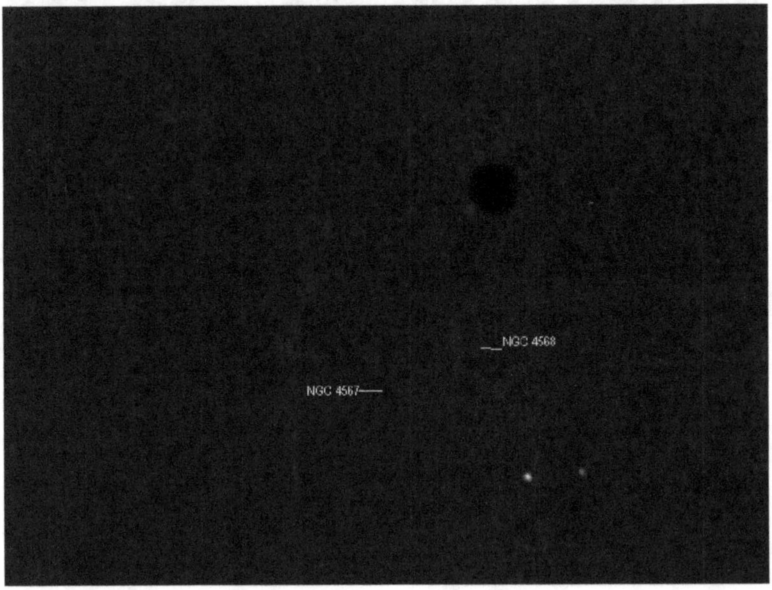

NGC 4567 and 4568 11″ SCT

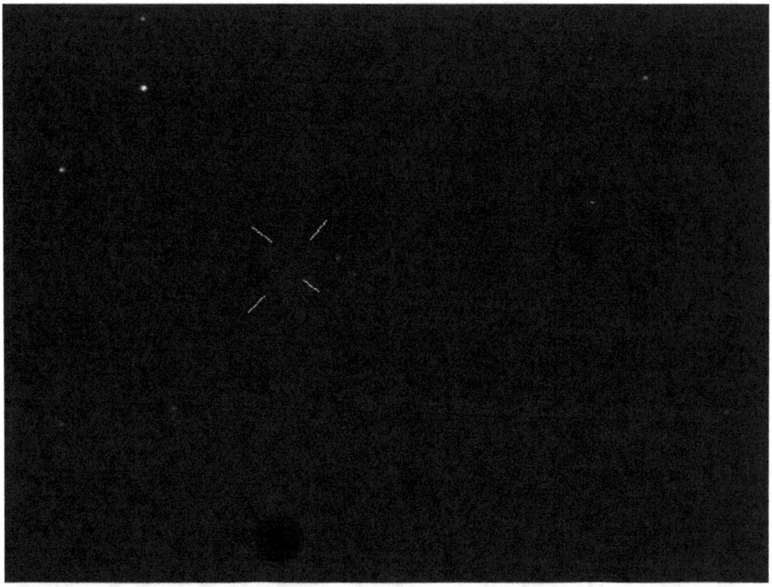

NGC 4567 and 4568 72mm Refractor

NGC 4699

This SABb galaxy has a very bright, stellar nucleus. The halo shows a very slight elongation around the nucleus.

NGC 4699 11″ SCT

NGC 4699 72mm Refractor

NGC 4762

This is an edge-on SB galaxy, with a compact core sandwiched between two 9th magnitude stars. How much of the extent of this spindle galaxy can you see?

NGC 4762 11″ SCT

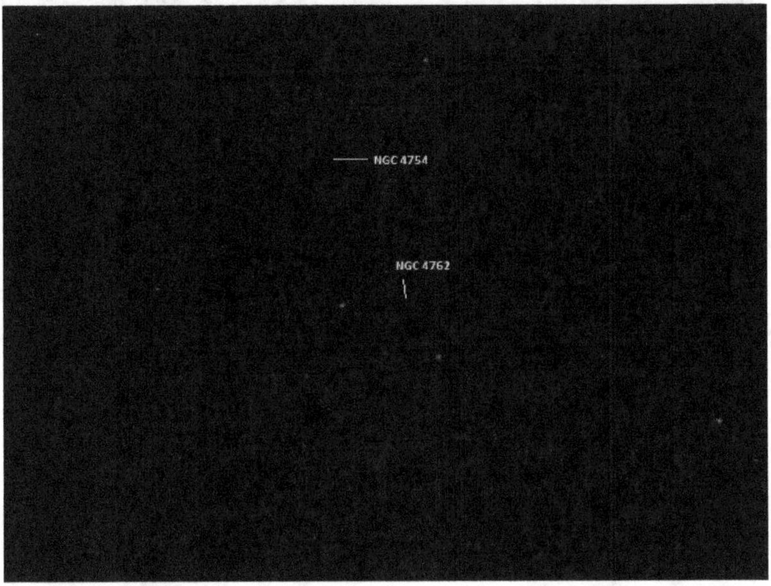

NGC 4762 72mm Refractor

NGC 5746

This is a SABb galaxy, presented edge-on to us. It has a bright core. Can you see the central bulge? If so, can you see the dark lane in it?

NGC 5746 11″ SCT

NGC 5746 72mm Refractor

Vulpecula

Objects	Type	Position	Size	SB*
Cr 399	OC	19h26′12.0″ +20 06 00	90′	
M 27	PN	19h59′36.1″ +22 43 13	402″	
NGC 6802	OC	19h30′35.0″ +20 15 48	5′	
NGC 6823	OC	19h43′10.0″ +23 18 00	7′	
NGC 6940	OC	20h34′32.3″ +28 17 00	25′	

Cr 399

This open cluster contains about 50 stars, the brightest of which make up the famous coat hanger shape. This cluster is also known as Brocchi's Cluster. As you can see from the pictures below, it requires a wide field of view to be fully appreciated.

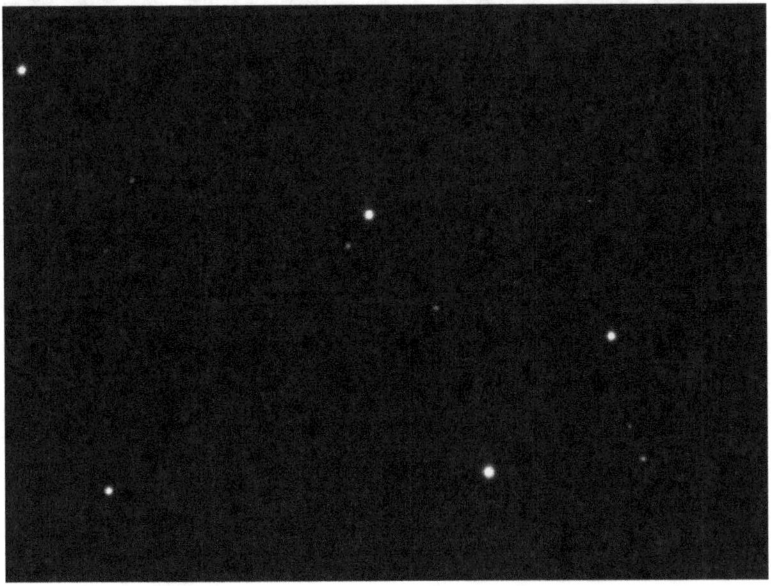

Cr 399 72mm Refractor

Trumpler Classification: III 3 m

M 27

This large, bright green planetary is better known as the Dumbbell Nebula. Can you make out the faint wisps between the ends of the hourglass shape? Use an UHC or an OIII filter.

M 27 11″ SCT

M 27 72mm Refractor

NGC 6802

This compact open cluster contains approximately 200 stars (the majority of which cannot be detected in amateur equipment) with a small range of brightness.

Trumpler Classification: I 1 m

NGC 6802 11″ SCT

NGC 6802 72mm Refractor

NGC 6823

This compact and dim open cluster contains almost 80 stars with a large range of brightness. Can you make out the nebulosity (NGC 6820)?

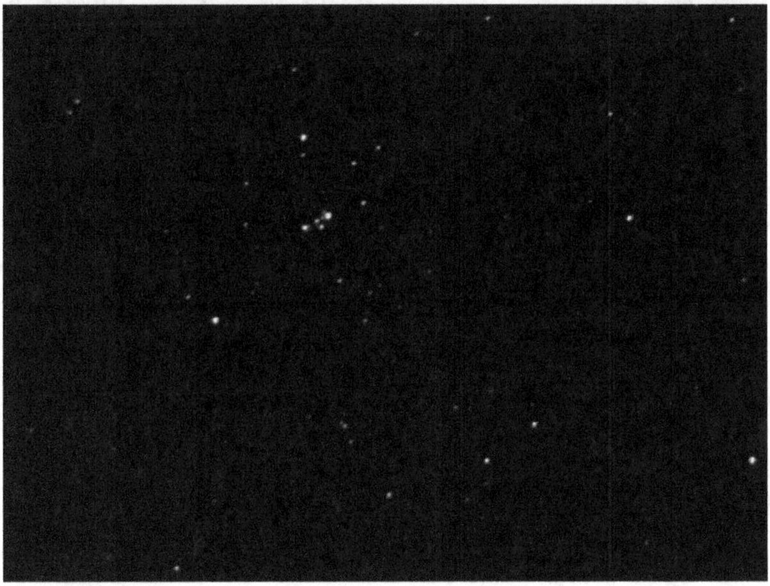

NGC 6823 11″ SCT

NGC 6823 72mm Refractor

Trumpler Classification: I 3 m n

NGC 6940

This large open cluster contains 170 stars with a moderate range of brightness. Note that it is not well detached from the field, but there is a significant concentration towards the center. Wide fields of view are required here.

Trumpler Classification: III 2 r

NGC 6940 11" SCT

NGC 6940 72mm Refractor

CHAPTER 5

OBSERVATION RECORDS

This page contains a sample sketching circle for you to photocopy and upon which to record your sketches of the objects in this book.

Object: ____________________

Date, Time, and Instrument: ____________________

Limiting Visual Magnitude: ____________________

Field of View: ____________________

Magnification: ____________________

Filters: ____________________

Seeing: ____________________

Weather: ____________________

Description:

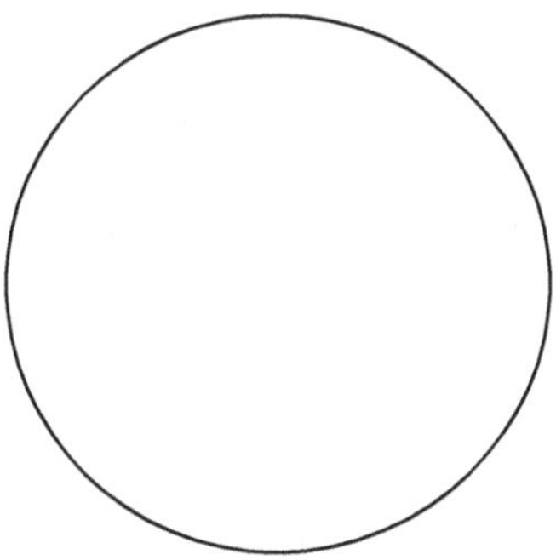

S. Rumistrzewicz, *A Visual Astronomer's Photographic Guide to the Deep Sky*, Astronomer's Pocket Field Guide, DOI 10.1007/978-1-4419-7242-2 _5,

RECOMMENDED TEXTS AND WEBSITES

The following are a list of recommended books on observing and several websites. Please note, that due to the nature of the internet, web addresses are subject to change, but were accurate at time of writing.

Recommended Texts

Allison, Mark, *Star Clusters and How to Observe Them* (Springer-Verlag, London, 2007)

Coe, Steven, *Nebulae and How to Observe Them* (Springer-Verlag, London, 2007)

Crogin & Bananno, *Uranometria Vol. 3* (Willmann Bell, Inc., USA, 2001)

Molisse, Rod, *The Urban Astronomer's Guide* (Springer-Verlag, London, 2006).

Steinicke, Wolfgang, and Richard Jakiel, *Galaxies and How to Observe Them* (Springer-Verlag, London, 2007)

Swanson, Michael, *The Nexstar User's Guide* (Springer-Verlag, London, 2004)

Thompson & Thompson, *Illustrated Guide to Astronomical Wonders* (O'Reilly)

Tirion, Rappaport, Remaklus, *Uranometria Vols. 1 & 2* (Willmann Bell, Inc., USA, 2001)

Useful Websites and Downloadable Software

Cartes du Ciel – free star chart software from the Internet by Patrick Chevalley

http://www.ap-i.net/skychart

NGC 891 – website containing observing lists and targets by Jim Cornmell

http://www.ngc891.com

Stellarium – free planetarium software from the Internet by Fabien Chéreau, Matthew Gates et al.

http://www.stellarium.org

Nexstar Resource Site – very useful website by Michael Swanson to accompany the book listed above, containing updates and free software for Nexstar telescopes

http://www.nexstarsite.com

Stargazer's Lounge – a wealth of information and friendly advice

http://www.stargazerslounge.com

Index

GPSR Compliance
The European Union's (EU) General Product Safety Regulation (GPSR) is a set of rules that requires consumer products to be safe and our obligations to ensure this.

If you have any concerns about our products, you can contact us on

ProductSafety@springernature.com

In case Publisher is established outside the EU, the EU authorized representative is:

Springer Nature Customer Service Center GmbH
Europaplatz 3
69115 Heidelberg, Germany

www.ingramcontent.com/pod-product-compliance
Lightning Source LLC
LaVergne TN
LVHW052217200926
848481LV00001B/138
9781441972415